Formation Evaluation

The EXLOG Series of Petroleum Geology and Engineering Handbooks

Formation Evaluation
Geological Procedures

Written and compiled by
EXLOG staff

Edited by Alun Whittaker

International Human Resources Development Corporation • Boston

ACKNOWLEDGMENTS FOR FIGURES

Figure 2-6 is reprinted by permission of the AAPG from Weaver, 1958.

Figures 2-7 are reprinted by permission of Elsevier Science Publishers after Karlsson et al., 1978.

Figures 2-16 and 2-17 are reprinted by permission of the Offshore Technology Conference from Clementz et al., 1979.

Figure 3-14 is reprinted by permission of the AAPG from Choquette and Pray, 1970.

© 1985 by EXLOG® .* All rights reserved. No part of this book may be used or reproduced in any manner whatsoever without written permission of the publisher except in the case of brief quotations embodied in critical articles and reviews. For information address: IHRDC, Publishers, 137 Newbury Street, Boston, MA 02116.

Library of Congress Cataloging in Publication Data

Main entry under title:

Formation evaluation.

(The EXLOG series of petroleum, geology, and engineering handbooks)
Bibliography: p.
Includes index.
 1. Rocks, Sedimentary. 2. Petrology. 3. Core drilling. 4. Petroleum—Geology.
I. Whittaker, Alun. II. EXLOG (Firm) III. Series.
QE471.F66 1985   622'.1828   85-2288
ISBN 0-88746-054-2
[90-277-1978-0 D. Reidel]

Printed in the United States of America

*EXLOG is a registered service mark of Exploration Logging Inc., a Baker Drilling Equipment Company.

TIME STRATIGRAPHIC COLUMN

AGE (m.y.)	ERA	PERIOD		DIVISION	MAJOR CLASSIFICATIONS			OROGENIC PHASES
0	CENOZOIC (Cz)	QUATERNARY (Q)		HOLOCENE (Ho)				
1				PLEISTOCENE (Ps)	Calabrian		Villafranchian	Pasadenian — LATE ALPINE
16		TERTIARY (T)	NEOGENE (Ng)	PLIOCENE (Pl)	Astain	Levantian / Dacian / Pontian / Meotian	Piacezian	Wallachian / Rhodanian
				MIOCENE (Mi)	Pannonian / Samartian / Vindobonian / Burdigalian / Aquitanian	Tortonian / Burdigalian / Aquitanian / Casselian	Helvetian	Attican / Styrian — MIDDLE ALPINE / Savian
25			PALAEOGENE (Pg)	OLIGOCENE (Ol)	Chattian / Rupelian / Lattorfian	Sannoisian	Tongrian	
42				EOCENE (E)	Priabonian / Lutetian / Bruxellian / Ypresian	Bartonian / Ledian	Auversian	Pyrenean
				PALAEOCENE (Pc)	Landenian / Montian / Danian	Sparnacian	Thanetian	
70 ± 2	MESOZOIC (Mz)	CRETACEOUS (K)	GULFIAN	UPPER (Ku)	Senonian	Maestrichtian / Campanian / Santonian / Coniacian / Argoumian / Ligerian	Dordonian	Laramide — EARLY ALPINE / Sub-Hercynian
86 ± 2					Turonian / Cenomanian / Albian	Vraconian		Austrian / Oregonian

TIME STRATIGRAPHIC COLUMN

TIME STRATIGRAPHIC COLUMN

Era	Period	Series	Stage	Substage	Orogeny
PALAEOZOIC (Pz)					
UPPER (PzU)	CARBONIFEROUS (C)	PENNSYLVANIAN	UPPER (Cu)	Stephanian	Asturian / Arbuckle / Young Wichita / Erzgebirgian / Old Wichita / Sudetian
				Westphalian	
				Namurian	
		MISSISSIPPIAN	LOWER (CL)	Uralian	
				Moscovian	
					Virgilian / Missourian / Desmoinsian / Atokan / Morrowan / Chesterian / Meramecian / Osagean / Kinderhookian
				Visean	Avonian
				Tournasian	
	DEVONIAN (D)		UPPER (Du)	Strunian	Bretonian / Arcadian
				Famennian	Bradfordian / Chautauguan / Senecan
				Frasnian	
			MIDDLE (DM)	Givetian	Mid-Devonian
				Couvinian	Erian
				Eifelian	
			LOWER (DM)	Emsian	Ardennian
				Siegenian	Ulsterian
				Gedinnian	
				Downtonian	
LOWER (PzL)	SILURIAN (S)	GOTHLANDIAN (G)		Ludlovian	Cayugan
				Wenlockian	Niagaran
				Llandoverian	Albion
				Salopian	
	ORDOVICIAN (O)		UPPER (Eu)	Ashgillian	Taconian
				Caradocian	Cincinnatian / Gamachian / Richmondian / Maysvillian / Edenian / Trentonian / Blackriveran / Chazyan
			MIDDLE (EM)	Lladeilian	Mohawkian
				Llanvirnan	Champlainian / Canadian / Trempealeauan / Franconian / Dresbachian
			LOWER (EL)	Arenigian	Sardinian
	CAMBRIAN (E)			Tremadocian	
PRECAMBRIAN (PE)		PROTEROZOIC	ARCHEAN		Many Major Orogenies
			ARCHAEOZOIC		

VARISCAN/HERCYNIAN — CALEDONIAN

Ages (Ma):
270 ± 10
350 ± 10
400 ± 10
500 ± 15
600 ± 20
3300 ±
4600 ±

CONTENTS

Frontispiece (Time Stratigraphic Column)   *v*

List of Illustrations   *xv*

Preface   *xvii*

1. INTRODUCTION
CUTTINGS RECOVERY   1
    DENSITY   1
    VISCOSITY   3
    GEL STRENGTH   6
    PIPE ROTATION   7
    PARTICLE SHAPE AND SIZE   8
    SUMMARY   9
CUTTINGS SAMPLING   9
    SHALE SHAKER   10
        Unwashed Sample   10
        Washed Sample   13
        Dried Sample   15
        Shipping Sample   16
    DESANDER AND DESILTER   17
    MUD   17
    COMBINED SAMPLE   17
CORE SAMPLING   18
ROCK CLASSIFICATION   18

2. DETRITAL ROCKS
CLASSIFICATION   21
    RUDACEOUS ROCKS   21
    ARENACEOUS ROCKS   22
    ARGILLACEOUS ROCKS   22
DESCRIPTION   24
    PARTICLE SIZE   25
    PARTICLE SHAPE   26
    SURFACE TEXTURE   27
    SORTING   27
    FABRIC   30
    STRUCTURE   30
    MINERALOGY   34
        Clastic   34
            Oligomict   34
            Polymict   34
        Resistate   35
        Hydrolysate   35
    INDURATION   39
        Clastic and Resistate   40
        Hydrolysate   40
    CEMENT   40
    MATRIX LITHOLOGY   42

    ACCESSORIES   42
        Minerals   42
        Moh's Scale of Hardness   47
        Fossils   47
    COLOR   47
    PETROLEUM SIGNIFICANCE   48
        Clay Diagenesis   49
        Abnormal Pore Pressure   54
        Source Potential   59
        Porosity   67
        Permeability   70
        Hydrocarbon Shows   73

3. CARBONATE ROCKS

CLASSIFICATION   75
    PETTIJOHN   76
    DUNHAM   76
    ARCHIE   76
        Matrix   78
        Visible   78
    FOLK   78
        Allochems   78
            Intraclasts   79
            Oolites   79
            Fossils   79
            Pellets   79
        Microcrystalline Calcite Ooze   80
        Sparry Calcite   80
        Limestone Categories   80
            Type I   80
            Type II   80
            Type III   81
            Type IV   81
        Limitations   81
    LEIGHTON AND PENDEXTER   83
    PLUMLEY   86
DESCRIPTION   86
    SAMPLE PREPARATION   87
        Cores   87
        Cuttings   88
    ROCK CLASS   90
        Sedimentary Carbonates   90
        Diagenetic Carbonates   91
    PARTICLE SIZE   91
    PARTICLE TYPE   93
        Carbonate minerals   93
            Aragonite   93
            Calcite   93
            Dolomite   94
        Origin   94

        Clastic Debris   95
            Bioclasts   95
            Colloclasts   96
            Carbonate Detritus   96
            Terrigenous Detritus   96
        Intact Material   97
            Pellets   97
            Oolites   97
            Bioliths   97
    PARTICLE SHAPE   98
    SURFACE TEXTURE   98
    SORTING   98
    FABRIC   99
    STRUCTURE   99
    MINERALOGY   100
        Recrystallization   100
        Dolomitization   101
        Solution   103
    INDURATION   104
    CEMENTATION   104
    ACCESSORIES   108
    COLOR   109
    PETROLEUM SIGNIFICANCE   109
        Porosity   109
            Primary Porosity   111
            Secondary Porosity   111
            Predepositional Porosity   113
            Depositional Porosity   113
            Eogenetic Porosity   113
            Mesogenetic Porosity   114
            Telogenetic Porosity   114
        Porosity Description   116
        Permeability   116

4. CHEMICAL ROCKS
INTRODUCTION   119
    NON-MARINE CARBONATES   119
    PRECIPITATION   119
    LACUSTRINE DEPOSITS   120
SILICEOUS ROCKS   120
    ORGANIC   120
    INORGANIC   121
FERRUGINOUS ROCKS   121
    CARBONATES   122
    SILICATES   122
    OXIDES AND HYDROXIDES   122
    SULFIDES   123
ALUMINOUS ROCKS   123
PHOSPHATIC ROCKS   123
SALINE ROCKS   124

                CALCIUM SALTS   129
                    Carbonates   129
                    Sulfates   129
                HALITE   133
                BITTERN SALTS   134
        CARBONACEOUS ROCKS   137
            CLASSIFICATION   137
                Peat   138
                Lignite   139
                Bituminous Coal   139
                Anthracite   140
            SOLID HYDROCARBONS   140
                Mineral Wax   140
                Asphaltites   141
                Pyrobitumens   141
5. IGNEOUS AND METAMORPHIC ROCKS
IGNEOUS ROCKS   143
        CLASSIFICATION   143
            Plutonic Series   143
            Volcanic Association   144
                Intrusive   144
                Extrusive   144
                Explosive   144
            Hypabyssal Rocks   144
        DESCRIPTION   145
            Silica Percentage   145
            Grain Size   147
            Texture   148
                Homeocrystalline   148
                Porphyritic   148
                Xenolithic   149
                Fragmental   149
                Grain Form   150
                Euhedral Granular   150
                Subhedral Granular   151
                Anhedral Granular   151
            Mineralogy   160
                Quartz   160
                Feldspars   160
                Feldspathoids   161
                Ferromagnesian Minerals   161
        ROCK NAME   154
METAMORPHIC ROCKS   154
        CLASSIFICATION   154
            Facies   154
            Parent   155
            Process   155
                Contact   161
                Autometamorphism   161

        Dynamic   161
        Regional   161
DESCRIPTION   161
    Texture   161
        Cataclastic   161
        Flaser   162
        Mylonitic   162
        Hornfelsic   162
        Granoblastic   162
        Slaty   162
        Phyllitic   163
        Schistose   163
        Gneissose   163
    Rock Name   163

APPENDIX A: FORMATION EVALUATION LOG SYMBOLS   165
References   171
Index   175

ILLUSTRATIONS

	Time Stratigraphic Column v-vii
1–1	Ideal Cuttings Recovery 2
1–2	Carrying Capacity of Drilling Fluids 4
1–3	Velocity Distribution in the Annulus 6
1–4	Effect of Pipe Rotation on Cuttings Recovery 7
1–5	Wentworth Particle and Standard Mesh Sizes 11
1–6	Sample Collection on "Rhumba"-Type Shale Shaker 12
1–7	Sample Collection on Doubledeck Shale Shaker 12
1–8	Microscope Samples 14
1–9	Examination of Washed Cuttings For Oil 15
1–10	Elementary Rock Classification 19
2–1	Wentworth Detrital Rock Terminology 23
2–2	Particle Shape: Roundness and Sphericity 26
2–3	Comparative Sorting 29
2–4	Mechanical Analysis for Size Distribution 29
2–5	Types of Bedding 31
2–6	Types and Amounts of Clay Minerals Commonly Found in Sediments 36
2–7	Mineralogical Variation in the Tertiary Section of Norweigian Well 2/11-1 37
2–8	C.E.C. Plot for the Tertiary of Norwegian Well 2/11-1 39
2–9	Cement and Matrix 41
2–10	Accessories Occurring in Sedimentary Rocks 43-46
2–11	Diagenesis and Metamorphism of Sheet Silicates 49
2–12	Theoretical Dewatering Curves 53
2–13	Theoretical Montmorillonite Composition Curves (Derived from Figure 2-12) 53
2–14	Effect of Abnormal Pore Pressures on Montmorillonite Composition 55
2–15	Overpressure Indication from C.E.C. 57
2–16	Van Krevelen Diagram Showing the Thermal Progression of Kerogens 61
2–17	Source Rock Evaluation by Pyrolysis 64
2–18	Example Pyroanalysis of Log Showing Common Events 66
2–19	Effect of Sorting on Porosity 69
2–20	Effect of Particle Alignment on Porosity and Permeability 69
3–1	Pettijohn/Dunham/Archie Carbonate Classification 75
3–2	Analogous Occurrence of Limestones and Detrital Rocks 77
3–3	Folk Carbonate Classification 78
3–4	Leighton and Pendexter Carbonate Classification 79
3–5	Energy Index Classification of Limestones for Sediment and Diagenetic Carbonates 84-85
3–6	Recommended Grain Size Classification 92

3-7	Carbonate Particle Type	95
3-8	Dolomitization by Evaporative Reflux	102
3-9	Spar Cement and Recrystallized Matrix	106
3-10	Drusy, Blocky and Rim Cement in Carbonates	107
3-11	Comparison of Porosity in Detrital and Carbonate Rocks	110
3-12	Genetic-Porosity Terms and Types	112
3-13	Porosity Types and Environments	115
3-14	Porosity Description	117
4-1	Evaporation of Seawater	125
4-2	"Ideal" Evaporite Section	127
4-3	Zechstein Evaporite Section	127
4-4	Characteristics of Gypsum and Anhydrite	130
4-5	Types of Gypsum-Anhydrite Conversion	132
4-6	Typical Salt Dome	134
4-7	Types of Coal	138
5-1	Classification by Color Separation	146
5-2	Igneous Grain Size Classification	147
5-3	Grain Form and Rock Texture	151
5-4	Igneous Rock Classification	156-157
5-5	Metamorphic Facies	158
5-6	Metamorphic Facies: Mineralogy and Occurrence	158
5-7	Metamorphic Parents and Products	159
5-8	Common Mineral Assemblages	160
5-9	Metamorphic Rock Classification	164

PREFACE

Petroleum Geology is a complex discipline, drawing upon data from many technologies. It is the function of Wellsite Geologists to integrate processed data produced prior to and during the drilling operation with their own geological observations. For this reason, it is necessary that geologists appreciate some of the technology, theory of measurement, and processing of this data in order to better assess and use them.

In the *Field Geologists's Training Guide* (Exlog, 1985) and *Mud Logging: Principles and Interpretations* (Exlog, 1985), an introduction is given to the scope of petroleum geology, and the techniques of hydrocarbon (oil and gas) logging as a reservoir evaluation tool. This handbook is intended to provide the Logging Geologist, and those training for a Consultant Wellsite Geologist position, with a review of geological techniques and classification systems. This will ensure the maximum development of communicable geological information.

Whether a geologist's work lies in this direction or in the more applied field of pressure evaluation, it is the application of geological insight to engineering problems that distinguishes the professional logging geologist in the field. This book will be of interest to and become a regular reference for all geologists.

1
INTRODUCTION

1.1　　　　　　　　　　CUTTINGS RECOVERY

In an ideal borehole and mud system, cuttings would be transported to surface with the same order and composition as they were cut, as in Figure 1-1. In reality, it is commonly observed that, no matter how sharp a drilling break or clearly defined a formation top on an electric log, the formation top as logged from cuttings will appear to be "transitional." (Note: not transitional in the sense that the cuttings show a varying composition, but that cuttings of a new type appear in increasing abundance as more hole is made.) The reason for this is that, in transit in the annulus, cuttings tend to travel at different net velocities, resulting in a relative "slippage" and sorting. That such a process can take place is demonstrated at surface. Mass sorting, a function of cuttings density and size, is seen to take place at the shale shaker and in the sieve while washing. A similar process is also seen to occur in centrifuge or hydroclone separation, which may be defined as having a specific "cut" or minimum particle size to be separated. A device designed to remove a particular "cut" of clay, silt or sand particles will in fact remove much smaller particles of denser materials such as barite. Thus a supposed size separation process is in fact a separation process controlled by particle mass and shape.

The complex of flow regimes through which cuttings pass between the formation and the flowline (including the various breaks in circulation for connections) results in a major sorting of cuttings. This will be affected by the relative densities of various cuttings and mud, the rheological properties of the mud and, since cross-sectional area affects the interrelationship, the size and shape of the cuttings. The combination of these factors causes highly variable and not always predictable results. Very little quantitative research has been performed, but the following examples illustrate the effects of changes in mud properties and flowrate on the recovery of particles of similar shape and density but different sizes. These are illustrated in Figure 1-2. The tests were performed by Williams and Bruce on a 500-ft test well using various sizes of aluminum disks.

1.2　　DENSITY　(Refer to Figure 1-2)

Comparison of mud (c) and mud (d) shows that increasing mud density increases the carrying capacity, thus delivering the maximum quantity of cuttings in the minimum time. (Mud circulation time for this test system is approximately four minutes.) With similar rheology, mud (d) recovers 75 percent of the total material in under two circulations, while mud (c) requires more than three to achieve a similar result. The reason for this is readily apparent. Decreased density of fluid relative to that of the carried particles decreases the weight of the particles in the fluid and therefore decreases the capacity of the fluid to both suspend and carry them.

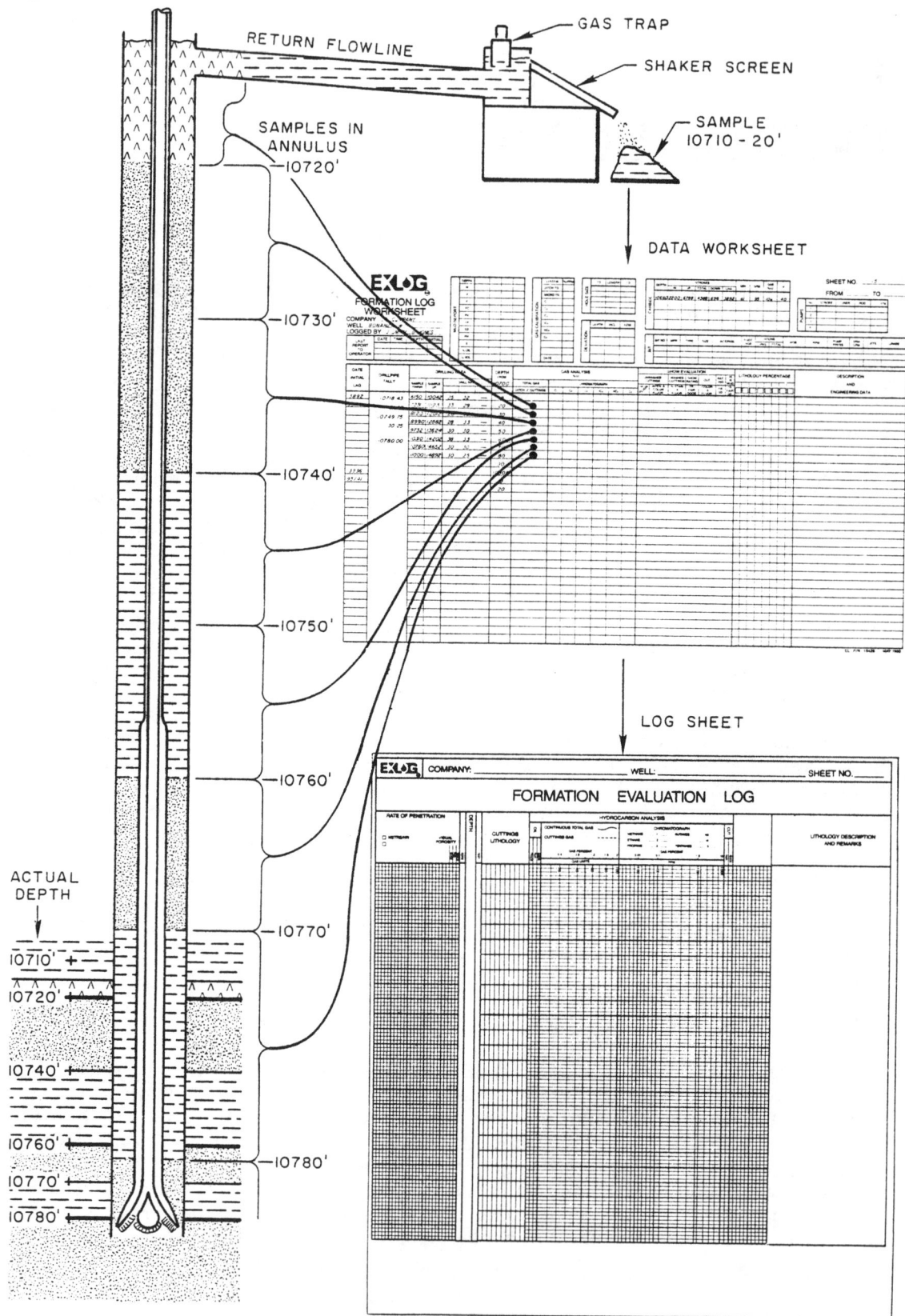

Figure 1-1. Ideal Cuttings Recovery

1.3 VISCOSITY (Refer to Figure 1-2)

Intuition suggests that increasing the viscosity would improve the mud's capability for carrying cuttings. However, comparison of (a), which is water, with muds (b) and (c) shows that increasing viscosity, even when accompanied by increased fluid density, actually decreases the efficiency of cuttings recovery. Mud (f) with a viscosity of 30 centipoise shows a recovery rate of the same order as water, but this appears to be a result of its 12.4-lb/gal density since the recovery rate is less than mud (d) with its lower density and viscosity. The reason appears to be that, at higher viscosities, the fluid is more likely to be in laminar flow which is less efficient for cuttings transport.

When a solids particle is carried by a fluid up an annulus, the particle moves upward with the fluid, but the effect of gravity on the denser particle retards the upward movement. The net upward velocity of the particle is the difference between the upward mud velocity and the downward slip velocity of the particle:

$$V_u = V_a - V_s$$

$$V_s = \frac{V_c}{1 + d/2\Delta} \tag{1-1}$$

$$V_c = R\sqrt{\frac{gd(\rho_p - \rho_f)}{\rho_f}}$$

where

V_u = cuttings net upward velocity

V_a = mud (bulk) annular velocity

V_c = calculated theoretical slip velocity

V_s = slip velocity corrected for velocity distribution and well effects

d = diameter of particle

Δ = hydraulic diameter (hole I.D. minus pipe O.D.)

g = gravitational constant

ρ_f = fluid density

ρ_p = particle bulk density

and

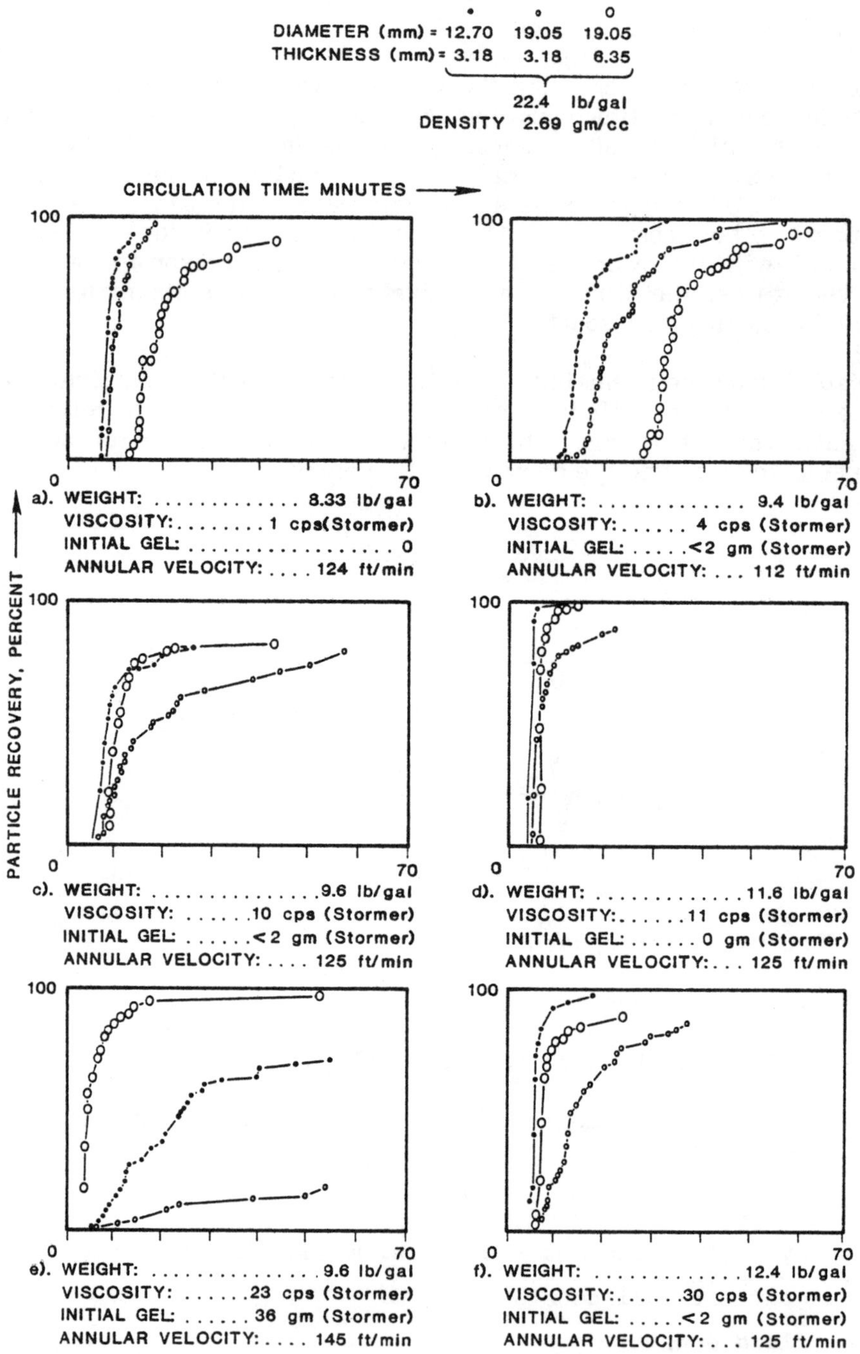

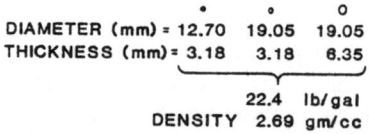

Figure 1-2. Carrying Capacity of Drilling Fluids

$$R = 1.35\sqrt{t/d} \quad \text{for flatwise fall of a disk}$$

$$= 18\sqrt{t/d} \quad \text{for edgewise fall of a disk}$$

$$= 1.74 \quad \text{for fall of a sphere}$$

where

t = thickness of disk

d = diameter of disk

As this difference becomes increasingly positive, particle recovery becomes more efficient. If it is negative, that is, if the downward slip velocity is greater than the upward mud velocity, particle recovery does not take place. However, not all of the fluid in the annulus is moving at the same velocity (see Figure 1-3). In turbulent flow the fluid elements move in countless eddies, swirls or "turbs." In laminar flow the fluid elements follow the streamlines or "laminae." The overall result is that the statistical vector average of all "turbs" at any point falls close to the average velocity of the fluid as a whole, while velocity distribution throughout the streamlines shows a much wider variation from the average.

The behavior of carried particles under the two flow types is also very different. Under turbulent flow the particles are carried in an even manner and, although disturbed by the turbulence, they tend to maintain the maximum surface area perpendicular to the mean direction of flow. As would be expected, mass sorting takes place, resulting in recovery in the following order: small, medium, large size particles (see Figure 1-2, a and b).

Under conditions of laminar flow the behavior of suspended particles varies according to their dimensions. In the case of the aluminum disks, the large disks moved up the annulus in a manner similar to turbulent flow. The medium and small disks were turned on edge and moved to the outer wall (and to the pipe when it was not rotated) and slipped down before beginning to rise again. Again, due to their greater mass, the medium disks tended to slip back further than the small. In some cases these disks were held against the borehole wall and not recovered, without increasing annular velocity. The result is recovery in the order large, small and medium sized particles, with notably poorer (less than 50 percent) total recovery of medium sized particles (see Figure 1-2e). Due to the overall mass sorting effect in the annulus, in denser muds (Figure 1-2, d and f) small and large particles were recovered at approximately the same rate, with medium sized particles less efficiently recovered — the turning of the particle being the most important factor (see paragraph 1.6) but particle mass being a secondary influence.

Due to the turning motion of the medium and small particles, the slip velocity Equation (1-1) which was empirically derived for turbulent flow cannot be applied with nonspherical particles in laminar flow due to the constant change in their orientation. Slip velocity, in addition to mud velocity, varies across the annulus.

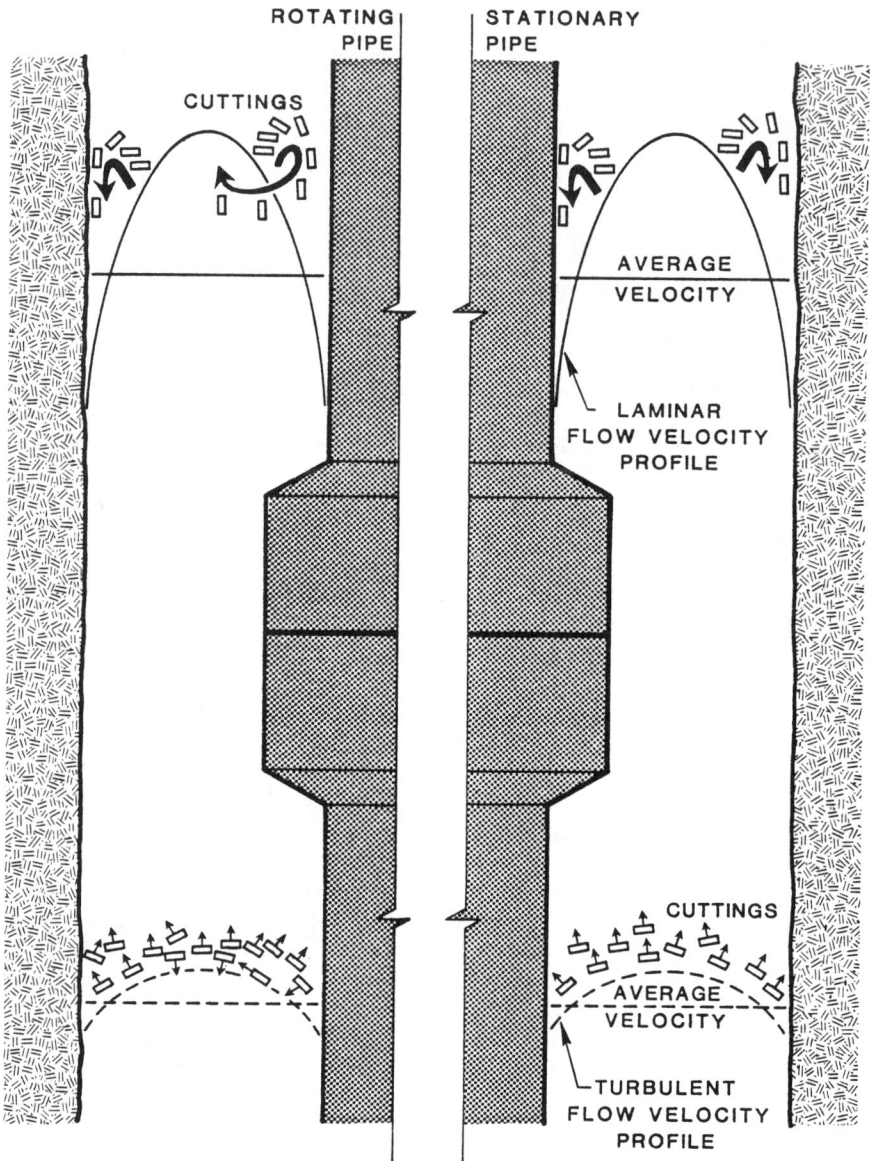

Figure 1-3. Velocity Distribution in the Annulus

1.4 GEL STRENGTH

The effect of gel strength is not readily obvious due to the common relationship between that quantity and viscosity. It is common to assume that a high-viscosity mud will also have a high gel strength. This is true if viscosity is due to a high percentage of well-dispersed high-yield clays such as "premium" bentonite. However, if the high viscosity results from a concentration of inert solid additives, drilling solids, poorly dispersed or low-yield "natural" (formation) clays, the gel strength remains low.

From the muds studied by Williams and Bruce, no specific conclusions can be drawn though it is generally concluded that increased gel strength inhibits the effect of mass and size sorting by restricting particle movement relative to the fluid. This is certainly true when circulation is stopped and highly gelled mud prevents fallback of cuttings (Figure 1-4). It is interesting to note that when high-viscosity high-gel muds were tested, the first particles arrived earlier than calculated from the (bulk or average) annular velocity of the mud, suggesting that a similar effect takes place when mud is being circulated. Thus some particles, even in laminar flow, were being held at the point of highest mud velocity such that their net velocity (mud velocity minus slip velocity) was higher than the average mud velocity.

1.5 PIPE ROTATION

All of the examples in Figure 1-2 were obtained with the pipe rotating in the hole. It was found that, without pipe rotation, particle recovery was severely reduced (Figure 1-4). This effect is most pronounced in high-viscosity and high-gel-strength muds. However, ranges of rotary speed between 35 and 150 rpm produced no appreciable change in particle recovery.

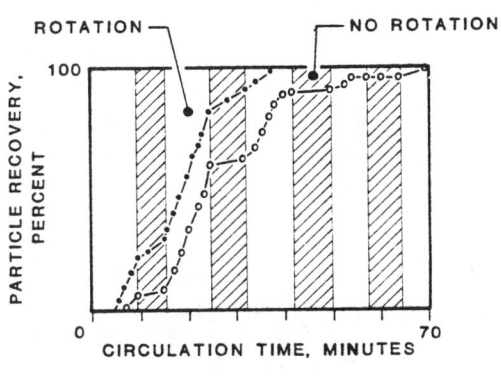

- LOW VISCOSITY/LOW GEL MUD
- HIGH VISCOSITY/HIGH GEL MUD

Figure 1-4. Effect of Pipe Rotation on Cuttings Recovery

When the pipe is rotating and the mud is in laminar flow, those small and medium sized particles that are carried on edge to the pipe are turned by the pipe and thrown back into the high-velocity portion of the annulus, thus aiding in particle recovery. This effect also occurs in turbulent flow, but is less noticeable (Figure 1-3).

1.6   PARTICLE SHAPE AND SIZE

For the purpose of this experiment, aluminum disks were used, hammered to give a roughened, irregular, cuttings-like surface. In terms of shape, size and density they would behave like very large cuttings or cavings fragments ranging from zero porosity sandstone to 8 percent porosity dolomite. They have a large area flattened shape more typical of cavings than cuttings. Despite this lack of similarity to commonly encountered cuttings, it is believed that the results of these experiments on a shallow (500 ft) experimental well are similar to the effects on real cuttings from deeper drilling wells.

The fact that the recovery of particles varied in laminar flow was found to be a function of cuttings shape as well as size and mass. For the three types of disk used:

Disk	Large	Medium	Small	
Diameter (d)	19.05	19.05	12.70	mm
Thickness (t)	6.35	3.18	3.18	mm
Density	2.69	2.69	2.69	g/cc
Volume	1.81	0.91	0.40	cc
Mass	4.87	2.44	1.08	g
Ratio t/d	0.33	0.17	0.25	
Flat section area	285.02	285.02	126.68	mm^2
Edge section area	120.97	60.58	40.39	mm^2

It was found that disks tend to travel with maximum cross-sectional area perpendicular to the direction of mudflow. Where the turning effect (due to velocity distribution) was strongest, in laminar flow the effect was in inverse proportion to the ratio of thickness to diameter. Thus the medium sized particles in this case having the minimum t/d of 0.17 were most affected, and the large sized particles with t/d of 0.33 were least affected.

Experiments with various other disks suggested that the critical thickness-to-diameter ratio was 0.3. Below this value disks would usually turn on edge in laminar flow; above it they would usually remain horizontal. It was also seen that disks with a thickness-to-diameter ratio of more than 0.78 would travel on edge at all times since, beyond this value, the cross-sectional area presented to the flow was greater when on edge than when horizontal.

It is important to note that most rock bit cuttings have "thickness-to-diameter" ratios less than or just greater than 0.3. Therefore, they will be strongly subject to the turning or "torque" effect when the mud is in laminar flow.

1.7 SUMMARY

In conclusion, it can be said that for ideal cuttings recovery:

- Mudweight should be maximized
- Viscosity should be minimized
- Gel strength should be reduced but not so low as to prevent cuttings suspension when circulation stops
- Annular velocity should be controlled to obtain turbulent flow
- The drillpipe should be rotated when circulating cuttings (that is, even when not drilling)

It is obvious that a drilling mud program designed to meet only these criteria would be otherwise unsuccessful. For example, excessive mudweight would reduce rate of penetration, too low viscosity would not supply the competence to carry the required high load of cuttings from the hole in fast drilling, and annular velocities sufficient to produce turbulent flow in all sections of the annulus would result in erosion and serious cavitation in the open hole section. The geologist's ideal requirements must be compromised with the practical requirements for economically and safely drilling and completing the well. It is therefore necessary for the geologist to understand and play a part in the overall drilling process so that mud properties and drilling practices may be optimized to provide satisfactory results for both himself and the engineer.

Since compromise conditions are the best that can be hoped for, the geologist must bear in mind the above factors and try to use them when evaluating cuttings. The most important point to remember is that the lag time, however determined, can be no more than a rough estimate of the arrival time of the <u>first</u> material from a drilling break. Some material may arrive before that time, and certainly will continue to arrive for some time (probably several circulations) after it. Although cuttings are the only direct indication of the formation being drilled, other indirect indicators, such as rate of penetration and drilling torque, may be more reliable indicators of formation changes. They should always be considered in an overall interpretation.

1.8 CUTTINGS SAMPLING

When using drill cuttings in order to determine formation lithology, representative sampling from all shale shaker screens, desander and desilter is essential. Procedures for correctly doing this and for correct washing and processing of samples are fully discussed in the <u>Field Geologist's Training Guide</u> and <u>Mud Logging: Principles and Interpretations</u> (EXLOG, 1985). Representative samples of rock cuttings from retrieval point (shale shaker, desander, etc.) and from the washed and sieved cuttings are of prime importance.

The geologist should always be aware of what size and type shale shaker screen is in use, when the screen is changed and when desilters or desanders are being used. By knowledge of the "cut" at each retrieval point and inspection of all samples and sieved samples of barite and drilling clays, it is possible to establish the presence and type of contaminants and drilling solids. This establishes a background against which future cuttings and formation detrital material may be judged.

Figure 1-5 indicates standard Wentworth particle and relative mesh sizes for shale shaker screens and sieves. In using the chart for comparison, remember that, unless the formation is unconsolidated or very poorly cemented, the particle size within the rock fabric will be less than the size of the recovered cuttings.

1.9 SHALE SHAKER

Take at least a pint of cuttings in addition to that needed to fill sample sacks. If the sample contains a large proportion of unconsolidated clay and the oil company wants this washed out of washed and dried samples, a larger quantity of cuttings will be required in order to provide reasonable quantities of washed and dried sample.

If the shaker is the "Rumba" type (Figure 1-6), take separate samples from all three screens. If it is the "double decker" type (Figure 1-7), take separate samples from top and bottom screens. After taking the sample, accumulated material on the shaker screen should be scraped off. If a board or bucket is being used to catch cuttings, it should be emptied after the sample is taken. Use of a board or buckets to catch cuttings may be necessary when there are few or poor sample returns. They are not a substitute for regular visits to the shale shaker and should not be required for normal returns.

Take the sample to the logging unit and use the following procedure.

1.10 Unwashed Sample

1. Place the cuttings in an 8-mesh sieve stacked on top of 80-mesh and 170-mesh sieves. Rinse lightly to remove drilling mud.

2. If the sample consists of unconsolidated clay, remove a small sample for microscopic examination.

3. Blend 100cc of sample with 600cc of water for thirty seconds. Let stand for a further thirty seconds and take a cuttings gas reading.

4. Inspect the water surface for oil droplets and petroleum ordor.

5. Report the amount and color of oil droplets. If sufficiently abundant, skim off a sample.

6. Set aside the blended sample for later examination (see paragraph 1.11).

7. If oil indications are seen or expected, place about 50 cc of unwashed sample in a dish and inspect under ultraviolet light for oil fluorescence.

8. If no fluorescence is seen, add 100 cc of water and stir the sample. Observe for fluorescent oil droplets "popping" to the water surface.

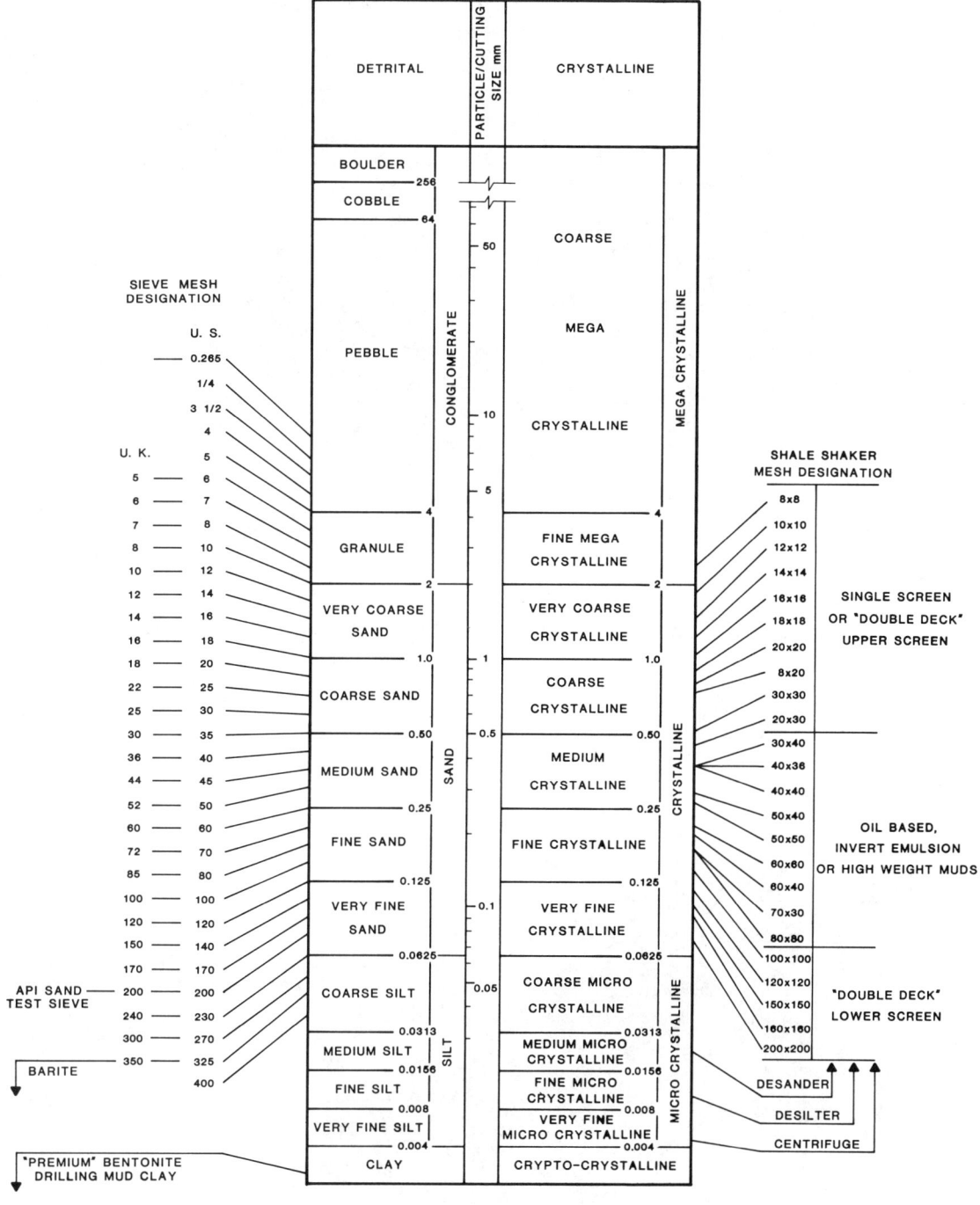

Figure 1-5. Wentworth Particle and Standard Mesh Sizes

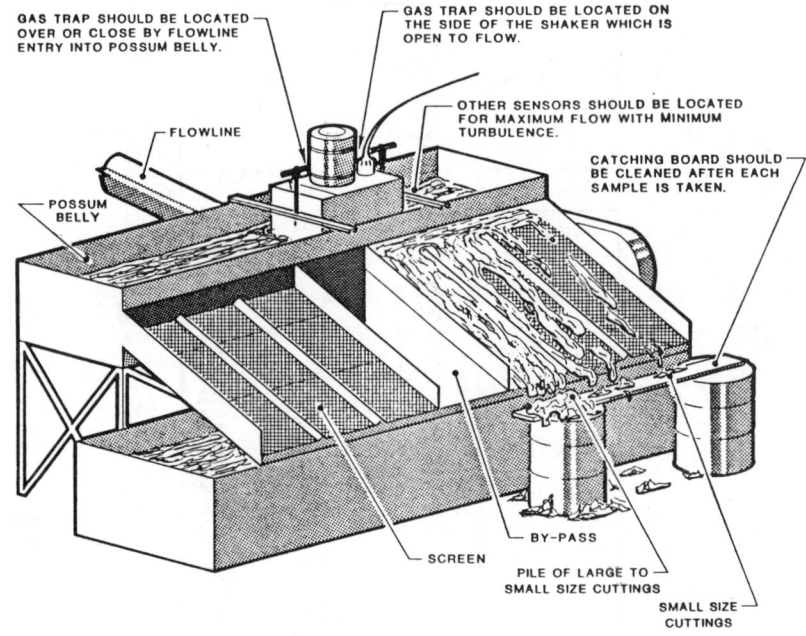

Figure 1-6. Sample Collection on "Rhumba"-Type Shale Shaker

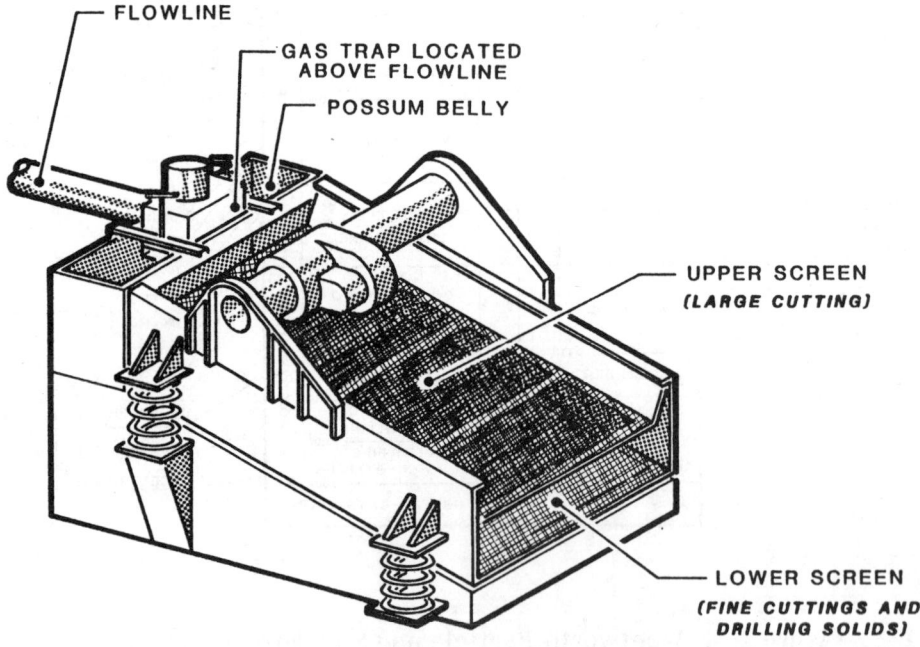

Figure 1-7. Sample Collection on Doubledeck Shale Shaker

9. Report the amount, color, intensity, and distribution of oil and fluorescence.

1.11 Washed Sample

1. If a "rinsed-only" sample is required, remove a sufficient amount to fill a dried sample envelope.

2. Manipulate unwashed sample with your fingers to estimate clay content and consistency.

3. Wash it through the 8-mesh sieve and take sample (probably cavings) for examination.

4. Wash the 8-mesh washings through the 80-mesh sieve and take sample for examination. Take care to prevent density separation of grains in the sieve.

5. Wash the 80-mesh washings through a 170- or 200-mesh (sand test kit) sieve and take sample for examination.

6. Decant the 170-mesh washings and take sample for examination.

7. Wash blender residue through a 170- or 200-mesh sieve. Take sample from sieve and decanted washings for examination.

8. Arrange the sample as shown in Figure 1-8 for inspection under the microscope.

9. Arrange a similar sample for inspection under ultraviolet light.

10. At the microscope, describe the various lithologies represented in the sample according to:

    rock type
    color
    hardness or induration
    grain size
    grain shape
    sorting
    luster
    cementation or matrix
    structure
    porosity
    accessories
    inclusions

11. Pick out those cuttings that have apparent oil staining. Place a representative selection onto a spot plate, one cutting per spot.

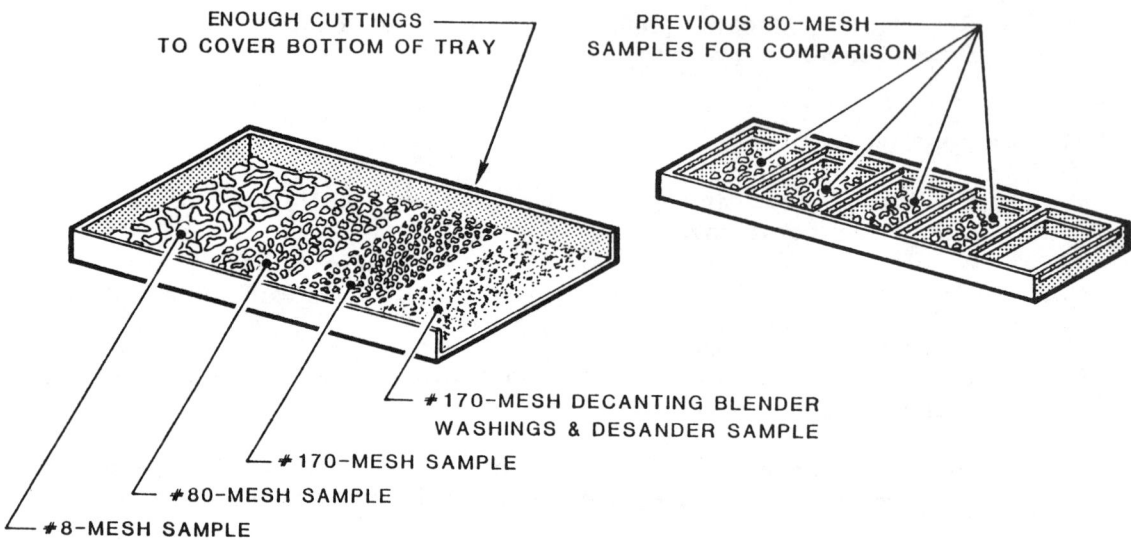

Figure 1-8. Microscope Samples

12. At the ultraviolet-light box, pick out those cuttings which fluoresce. Place a representative selection of these on a spot plate, one cutting per spot.

13. Take the spot plate from step 11 to the ultraviolet-light box and those from step 12 to the microscope. Confirm that apparent oil staining and fluorescence are present in both sets of sample cuttings (see Figure 1-9).

14. If not, identify the cause — such as mineral fluorescence, contamination with grease, pipe dope or plastic materials, residual "dead" oil stain, etc.

15. Report the amount, type, color and distribution of the oil stain and fluorescence.

16. Test fluorescent and oil-stained cuttings with solvent. Use a spot plate with one cutting per spot and one spot empty. Add solvent to each spot and use the empty spot as a reference.

17. Report the rate and manner that the oil fluorescence cuts (that is, disperses into the solvent); for example: instant, fast or slow, uniform, or streaming.

18. Report the color and intensity of the cut fluorescence and the natural color of the dissolved oil.

OIL TESTS - WASHED CUTTINGS

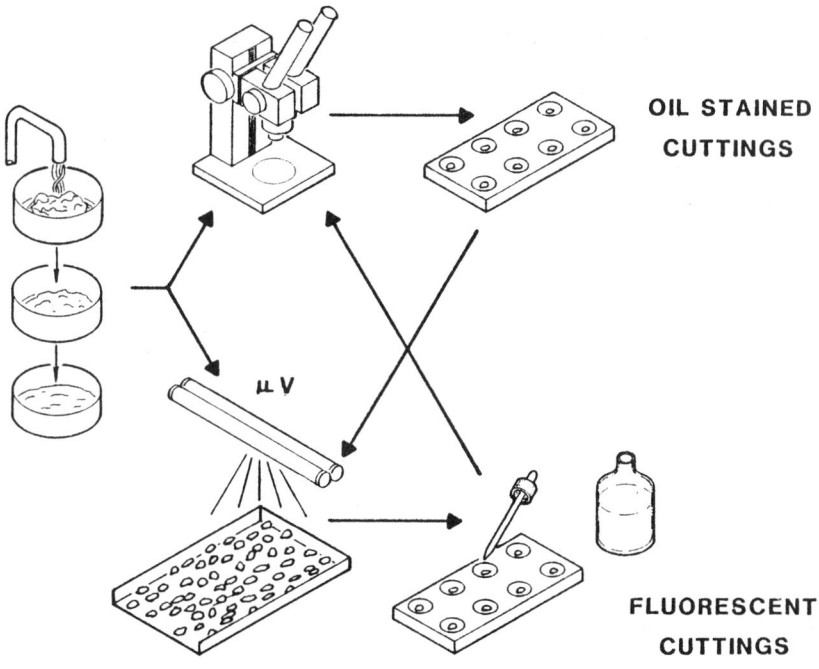

Figure 1-9. Examination of Washed Cuttings for Oil

1.12 Dried Sample

1. Rinse a larger quantity of sample through the 8-mesh into the 80-mesh sieve. Place on a sample tray in the oven.

2. Remove a small portion of dried sample and inspect under the microscope. Observe size, shape and distribution of porosity, surface texture and grain luster or coating.

3. If washed sample fluoresced but did not cut, repeat tests on dried cutting. Test solvent reaction of crushed or acidized cuttings.

4. Save bulk of dried sample in a labeled sample envelope.

NOTE

When drilling through an unconsolidated clay section, the oil company geologist and the logging geologist should confer as to the degree of washing required for "washed-and-dried" sample. A rinsed clay sample, upon drying, will become a useless, uninformative, clay "brick." A sample with all of the clay washed out may lead to false conclusions if reviewed later and taken to be a whole sample.

1.13 Shipping Sample

Cuttings samples are collected at a specified sample interval and shipped unprocessed from the wellsite. These may be sent to

- A Sample-Splitting Service for washing and repackaging for the oil company geologists
- Micropaleontologists or palynologists
- Geochemists
- Partners in the well
- Other oil companies for trade
- A government agency, e.g. Geological Survey or Ministry of Petroleum
- Storage for future reference or research

The number of samples and the sampling frequency are normally specified by the Logging Program or instructions prior to the well. The sampling requirements should not be so great as to prevent the logging geologist from carrying out his other duties. If they are, the client must be informed and, if possible, arrangements made to provide a sample catcher. The logging geologist has a number of professional responsibilities, and these should not be neglected at the expense of catching large volumes of sample.

When drilling is slow, the logging geologist should catch samples for his own evaluation over shorter depth intervals (approximately every 15 minutes) than the sample interval specified for the shipping samples. To ensure good representative sample for both purposes, each time a sample is taken for the logging unit, a small increment should be added to the sample sack for the current interval, and the shaker screen should be cleaned off. The 80-mesh portion of the unit sample should be set aside to be used in the dried sample for the interval. When the interval has been completed, the sack should be closed up and a new sack started.

It is important that the sack contain samples representative of the entire interval shown on its label (e.g. the 7980 - 7990 sample should contain samples from the whole ten feet). When drilling is slow, if cuttings are collected only at the end of the interval (e.g. at 7990), the sample may be representative only of the last few inches drilled. If a narrow band within a sample interval occurs with a unique lithology, both an interval sample and a special sample from the narrow band should be taken.

Shipping samples are commonly collected and placed unwashed into cloth sacks. The well name and interval (both top and bottom depth) are marked in waterproof ink on an attached tag. It is a good precaution, against the tag being torn off, to mark the interval on the bag too. If the cuttings are liberally coated with mud, it is a good idea to rinse them <u>lightly</u> with clean water before sacking. If possible, the sacks should be hung for a period to drip and air dry prior to shipping. Instructions from the oil company should be followed on this.

The sacks are normally shipped in large boxes or crates or in large burlap or hessian sacks. Before shipping, all sample bag labels should be checked to ensure they are legible and undamaged. Illegible labels or torn sample bags should be replaced. When shipping in boxes, samples should be packed in correct depth order. In sacks the samples should be tied in bundles of one hundred feet each. This will earn the gratitude of anyone who receives and unpacks the samples.

The shipping container should be clearly labeled with the well name and the total depth interval contained, and a record of these details and the date, destination and means of shipment kept in the unit. On a "tight" hole the oil company may not wish the well name or depths to be shown on labels. In this case, some numbering code should be mutually prearranged.

If geochemical and palynology samples require special treatment, instructions will be given. These samples will not be washed or even rinsed, and for geochemistry a sample of mud may be requested with the cuttings. Palynology samples may be requested unwashed but dried, and special arrangements may be necessary to set up a drier outside the unit. Both types are normally sealed, either in cans or plastic, and treated with a bacteriocide prior to shipping.

1.14 DESANDER AND DESILTER

Take a sample of effluent and process it in the same way as blender residue.

1.15 MUD

Take a mud sample after the shaker and process it in the same way as unwashed cuttings. A blender test for gas, oil and solid residue should be made. Inspection under ultraviolet light of fresh mud and mud diluted with water will show the presence of free oil.

In addition, in actual or potential pay zones, a sample of mud filtrate must be tested for salinity at least every fifty feet.

1.16 COMBINED SAMPLE

From a combination of the evaluations of each of the above derived samples, it is possible to reconstruct an accurate picture of the total rock material being returned from the borehole. Further evaluation and consideration are required to convert this evidence to an estimate of the actual rock type and structure in place.

Where there are neither payzones nor problems of evaluation, it may be considered unnecessary to follow this time-consuming procedure completely for every sample, especially when high rates of penetration limit the time available to be spent on each sample. Nevertheless, run the complete test periodically (each 100 feet or 2 hours, minimum). This will:

- Establish the nature and quantity of drilling solids
- Detect the presence of unanticipated fine, unconsolidated material

- Better establish the type and quantity of fine fractions and (possibly characteristic) detrital material
- Assist in estimates of grain size and sorting
- Provide possibly useful samples for micropaleontology
- Locate marginal and residual hydrocarbon accumulations

1.17 CORE SAMPLING

Instructions for sampling and packaging cores are included in the <u>Field Geologist's Training Guide</u> (EXLOG, 1985).

Samples of core material are excellent for describing and testing (oil evaluation, bulk density etc.) since they represent comparatively undisturbed and uncontaminated rock material and are not subject to the separation and/or mixing in the mudstream experienced by cuttings.

The whole core allows you to inspect the structure and macroscopic features which may be undetectable or only suspected in cuttings. Even if a core description log is not requested of the logging geologist, close inspection and detailed description of a core should always be made independently of the oil company geologist (with his permission).

1.18 ROCK CLASSIFICATION

There are many different methods of rock type classification. They may be based upon mineralogy, facies or environmental considerations. For the petroleum geologist, the classifications shown in Figure 1-10 prove most useful since they are based solely upon characteristics discernible at the wellsite from examination of well cuttings. However, not all items of this classification are of equal importance in petroleum geology.

It is assumed that the reader of this manual is a graduate geologist. For these reasons the rest of this manual discusses the various rock types only inasmuch as their occurrence, identification and description pertain to drill cuttings at the wellsite.

Sedimentary		
Detrital	Psephites/ Rudites	gravel conglomerate breccia tillite coarse-grained clastics
	Psammites/ Arenites	sand sandstone silt siltstone greywacke arkose fine-grained clastics
	Pelites/ Argillites	mud clay shale mudstone
Chemical Organic	Carbonates	limestone shell sand lime mud dolomite marl
	Siliceous	diatomite radiolarite chert flint
	Ferruginous	carbonates silicates oxides sulfides
	Aluminous	bauxite laterite
	Phosphatic	phosphorite
	Saline	rock salt halite potash gypsum anhydrite
	Carbonaceous	coal lignite anthracite bitumen
Igneous		
Volcanic association Plutonic series	———	———
Metamorphic		
Sedimentary origin		pelitic psammitic semi-pelitic calcareous calc-silicate
Igneous origin		basic acidic

Note: Certain of the rock names in lowercase above, e.g. gravel, shell sand, etc., are not recommended usage as will be outlined in paragraphs 2.1 and 3.30. They are included here as examples of common rock names used for members of each class.

Figure 1-10. Elementary Rock Classification

2
DETRITAL ROCKS

2.1 CLASSIFICATION

Detrital sediments consist of the broken, weathered, and transported fragments of preexistent rocks. During the processes of diagenesis and induration, further chemical and mineralogical changes may take place but these are not sufficient to alter the essential detrital character. Detrital rocks are sometimes referred to as terrigenous rocks. This term is used to specify the inorganic nature and physical origin of the rocks in contrast to the organically derived carbonates and chemically precipitated rocks. Literally, terrigenous means "from the land-mass", that is, from outside the depositional basin. The term "detrital" is a broader term encompassing material derived from both within and outside the basin.

Detrital rocks may be subdivided on a basis of mineralogy.

- Clastic: containing fragments of parent rocks and minerals with no newly formed components

- Resistate: containing fragments of inherited quartz and other detrital minerals resistant to chemical weathering

- Hydrolysate: containing transported clay minerals and other weathered and transported products of the parent rock

Although indicative of the nature of the host rock and the means and distance of transport, this classification is not generally acceptable since rocks may consist of a mixture of clastic, resistate and hydrolysate components. The simpler Wentworth Classification, which is based upon a single physical property — the mean grain size — is more applicable in the field.

- Psephites or Rudites: grain size discernible to the naked eye
- Psammites or Arenites: grain size discernible with hand lens or microscope
- Pelites, Lutites or Argillites: grain size indiscernible in the field

The terms Psephite, Psammite and Pelite and the terms Rudite, Arenite and Argillite are synonymous in definition. Unfortunately, some British geologists have taken the former terms to refer to metamorphic equivalent rocks. For this reason the latter terms will be used in this manual. This is also in common with most common modern usage.

2.2 RUDACEOUS ROCKS

The rudaceous or coarse detrital rocks, usually clastic, consist of particle sizes upward of 2 millimetres, usually larger. They are classified into three major types:

- Conglomerate: characterized by rounded particles bound by a finer material, usually silica, calcium carbonate or ferruginous minerals

- Breccia: characterized by angular, irregular fragments often in a sandy or finer-grained matrix

- Tillite: having fewer large fragments irregularly distributed in an unbedded fine-grained matrix. Glacial in origin

2.3 ARENACEOUS ROCKS

The arenites (or medium-grained clastic or resistate sediments) consist of particles of between 2.0 and 0.0625 millimetres. They are normally referred to as sand or sandstone. It is common for the term sand to be taken to mean quartz-sand. Arenaceous rocks may be formed from other mineral components, and confusion may be avoided by the addition of an appropriate prefix or by use of a specialized rock term:

- Orthoquartzite (Quartzarenite): quartz sandstones with quartz cement. Quartz constitutes more than 75% of the rock. Often well-rounded and sorted and often current bedded. Stable, mature deposits usually indicative of high-energy shallow sea margin environments.

- Greywacke (Litharenite): badly sorted and incompletely weathered fragments of unstable minerals and rock material in a finer grained matrix of similar composition. Quartz constitutes less than 75% of the rock, and lithic rock fragments are more common than feldspars. Indicative of rapid deposition (e.g., turbidity currents).

- Arkose: coarse fragments of quartz and feldspar, commonly in a calcitic or ferruginous cement. Quartz constitutes less than 75% of the rock and feldspars are more common than lithic rock fragments. Commonly terrestrial, indicative of rapid erosion and associated with conglomerates of similar parentage.

For purposes of cuttings description at the wellsite, using such specialized terms may be risky since they may imply environmental conclusions beyond a simple rock identification. For the logging geologist, the terms Sand or Sandstone should be used to specify the mean grain size of the sediment or rock. Identification of the grain mineralogy is added as a qualifier in the description, and the terms above are added, only as a final point, if conclusive evidence is seen (this may only be possible in a core). See Figure 2-1.

2.4 ARGILLACEOUS ROCKS

Argillites are the finest grained detrital rock with a grain size less than .0625 millimetre. The major constituents of these rocks are the newly-formed clay minerals resulting from the decomposition and hydration of the less stable minerals of the parent rock. Other new minerals which may be present include hydroxides, hydrous micas, sericitic and chloritic minerals generally. The third constituent of argillaceous rocks is silt, clay-sized or larger particles of inherited parent rock mineral, predominantly quartz with minor feldspar and mica.

PARTICLE			SIZE LIMIT (mm)		UNCONSOLIDATED		CONSOLIDATED	
			LOWER	UPPER				
RUDITES/ PSEPHITES	Boulder		256.0	—	Boulder	-Gravel	Boulder	-Conglomerate
	Cobble		64.0	256.0	Cobble	-Scree	Cobble	-Breccia
	Pebble		4.0	64.0	Pebble	-Till	Pebble	-Tillite
	Granule		2.0	4.0	Granule		Granule	
ARENITES/PSAMMITES	Very Coarse	Sand	1.0	2.0	Very Coarse	-Sand	Very Coarse	-Sandstone
	Coarse		0.5	1.0	Coarse		Coarse	
	Medium		0.25	0.5	Medium		Medium	(-Greywacke)
	Fine		0.125	0.25	Fine		Fine	(-Arkose)
	Very Fine		0.0625	0.125	Very Fine		Very Fine	
ARGILLITES /PELITES	Coarse	Silt	0.0313	0.0625	Silt		Siltstone	
	Medium		0.0156	0.0313	Mud		Mudstone	
	Fine		0.0078	0.0156			Claystone	
	Very Fine	Clay	0.0039	0.0078	Clay		Shale	
			—	0.0039				

Figure 2-1. Wentworth Detrital Rock Terminology

There is much controversy in geology, and specifically in petroleum geology, regarding the composition and nomenclature of argillaceous sediments. In the oil industry it is common to refer to all such sediments as "shale", ignoring the one characteristic, fissility, which the term demands. It is also commonly stated that, though silt may occur as a component in the matrix of rudaceous and arenaceous rocks, and as an often major component in argillaceous rocks, no rock of total silt composition, a true siltstone, will ever exist. Consideration of the environment of deposition of such a sediment would appear to support the contention that siltstones are rare, but they do occur in aeolian, fluvial overbank and in other low flow regime environments.

While all classifications and nomenclature have differing virtues and limitations, the following rock and sediment names have been chosen as best serving the purpose of wellsite description (Figure 2-1).

- Clay and Claystone: A structureless mass consisting predominately of clay minerals. Increasing depth and compaction will result in a change from a soft plastic unconsolidated material to a firm, hard, blocky rock.

- Mud and Mudstone: A clay or claystone containing a significant proportion of fine-grained or silt-sized material other than clay minerals. This will normally be quartz or feldspar but may be some other recognizable mineral. Mudstones containing significant quantities of calcite should be identified as calcitic mudstones. The term "Marl" is ambiguous and must be avoided unless its use is specifically requested by an oil company. In that case, ask the oil company to provide a definition of the term. This will vary greatly geographically and between companies.

- Shale: A mudstone or claystone exhibiting fissility, a finely laminated structure showing strong parallelism.

The mineralogical and chemical composition of an argillaceous rock at any point in time reflects a delicate equilibrium state of the clay mineral, organic matter and pore water phases. Study of this equilibrium composition may yield data leading to conclusions about

1) the mineral parent rock type
2) the organic parent type
3) the environments of weathering, transport and sedimentation
4) the diagenetic history of the rock
5) the current maturity of the formation, indicating the likelihood of encountering abnormal pressures or migrated hydrocarbons

NOTE

A conclusive test to distinguish claystone from mudstone is to grind a well-washed cutting between your teeth. The greater hardness of quartz and feldspar particles will give a "grittiness" to mudstone which is not felt with claystone. (The test may be duplicated using two knife blades as substitutes for teeth but this is more difficult to interpret.)

The term "siltstone," avoiding the argument of whether a pure example of such a rock can exist, is reserved for arenaceous rock, finer grained than a "very fine grained sandstone" but with visibly identifiable grains under the microscope.

2.5 DESCRIPTION

A rock description has two major functions:

1. It allows the reader to understand the structure and components of the rock and to draw valid conclusions as to the source and environment of deposition and subsequent history.

2. It allows another geologist to recognize the rock when he sees it (or determine whether he has seen it before).

For these reasons a rock description must be graphic and must include all observable features. Interpretive remarks may be added but must never replace the actual observation. Descriptions, while concise, must not be so cryptic as to allow false conclusions to be drawn. Unidentifiable features, e.g. detrital mineral or colored staining, should be described in detail even if their identity and significance are unknown to you.

NOTE

A standard abbreviation list of the rock names and descriptive terms is contained in Appendix A of <u>The Field Geologists Training Guide</u>, (EXLOG, 1985).

2.6 PARTICLE SIZE

Grain size determination from drill bit cuttings should follow a disciplined procedure in order to obtain an accurate overall estimate:

1. Size of individual grains

2. Mean size of grains in individual cutting

3. Mean size of grains in all cuttings (of the same lithology)

Unless the formation is unconsolidated, grain size determination using sieves or a micrometer ocular is not feasible. Therefore, an accurate visual estimate using a Sand Grain Folder and the above sequence should be used. Report the weighted average result taking into account the amount present of each grain size; but where the largest grains present are much larger than the weighted average, this maximum size should also be reported. Similarly, when the range of grain sizes is so large as to make a weighted average meaningless, note the minimum and maximum of the range.

In rudaceous rocks it is common for the matrix to consist of a fine-grained arenaceous sediment. Determine the grain size separately and phrase the description to express this separation, rather than showing it as a gradation in grain size from coarse to very fine.

In argillaceous sediments, take care not to mistake fine, disseminated, lustrous minerals for granularity. For example, a claystone containing microcrystalline pyrite or sericite will exhibit a surface "sparkle" reminiscent of a siltstone or fine sandstone.

2.7 PARTICLE SHAPE

There are an infinite number of variations of particle shape. For practical wellsite description these may be reduced to the functions of roundness and sphericity (see Figure 2-2). In addition to this, descriptive terms may be used to supplement, but not replace, the roundness and sphericity reported. For example:

Sharp	Disk	Faceted	Irregular
Angular	Elongate	Bladed	Fibrous
Flat	Rod	Tabloid	
Platy	Conchoidal	Blocky	

Grain shape is of critical importance in cuttings evaluation since it gives strong evidence of the distance and mode of transport, e.g., very angular/close to source, well-rounded/possibly aeolian, etc. In addition, grain shape is an important influence upon porosity and permeability — the two most important factors in reservoir evaluation. Increased roundness and sphericity will increase both porosity and permeability.

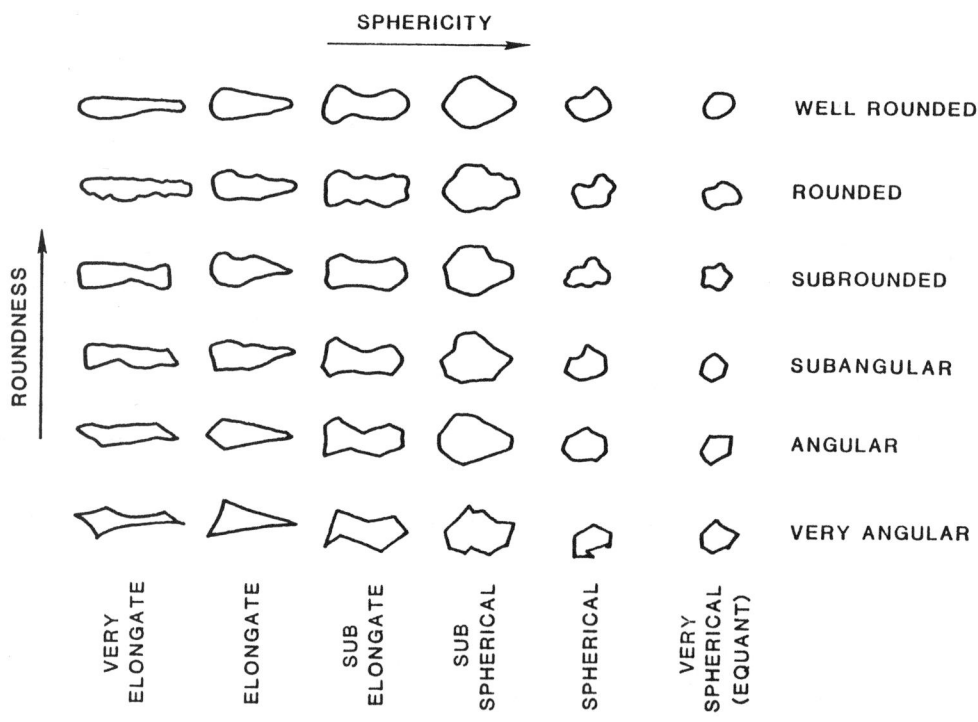

Figure 2-2. Particle Shape: Roundness and Sphericity

2.8 SURFACE TEXTURE

This term is used here to describe fine surface features of the particles — what might be called "relief" or grain "micro-shape." This is often described as "luster" since the texture may be so fine as to be observable only by its effect on reflected light. Observe the texture with the naked eye and the microscope and with the wet and dry sample. Rotating the sample tray relative to the light source also assists in detecting texture. Some common terms in use are:

- Coated, sooty: Precipitated or accretionary material on the surface of the particle, of insufficient thickness to develop a strongly visible color or "presence" (see Accessories, paragraph 2.23)
- Vitreous, glassy, faceted: Clear, shiny, fresh mineral appearance
- Silky, pearly, polished: Lightly etched or scoured
- Frosted, dull, etched: Deeply etched or scoured. Usually translucent white and roughened surface may be visible under high power
- Pitted: Solution or impact pits, often of pinpoint size
- Striated: Parallel abrasion lines or scratches
- Greasy, waxy: Slick, smooth, slippery appearance due to coating or grain mineralogy (e.g., hematite). Do not use the term "oily" which may be misunderstood

Use combinations of the above where applicable. Texture may be very important in determining the rock source, transport and history. It also assists in characterizing the rock for later recognition and correlation.

Although argillaceous rocks have no visible grains, the cuttings themselves have an observable texture. This may be indicative of mineralogy, accessories, hydrocarbon content, compaction history, or fracturing. Common terms used to describe pelitic rock textures are:

Earthy	Resinous
Waxy	Soapy
Greasy	Silky
Velvety	Sooty

2.9 SORTING

The degree of sorting of clastic rock is critical in determining its porosity and permeability. Unfortunately, it is also among the most difficult and subjective assessments made by the wellsite geologist, and normally extends no further than:

Extremely well sorted	
Very well sorted	Monomodal
Well sorted	
Moderately sorted	
Poorly sorted	Polymodal
Very poorly sorted	

No absolute standards are available to define degree of sorting other than typical examples such as those included in the Sand Grain Folder or in Figure 2-3. Plotting of "Mechanical Analysis Curves" (Figure 2-4) can give some quantitative aspect to sorting, but this is not normally practical at the wellsite. Numerical estimates of sorting quality and distribution may be computed using

$$M = \frac{d_{84} + d_{16}}{2}$$

$$M_d = d_{50}$$

$$C = \frac{d_{40}}{d_{90}}$$

$$\sigma_\phi = \frac{\log_2\left(\frac{d_{16}}{d_{84}}\right)}{2}$$

$$\alpha_\phi = \frac{\log_2\left(\frac{d_{50}^2}{d_{16}d_{84}}\right)}{\log_2\left(\frac{d_{16}}{d_{84}}\right)}$$

where:

    M = mean diameter (mm)
    M_d = median diameter (mm)
    C = uniformity coefficient
    σ_ϕ = sorting coefficient (standard deviation)
    α_ϕ = skewness coefficient
    d = particle diameter (mm)

d_{16}, d_{84}, d_{50}, etc. = d_n

where:

    n = weight percentage of samples having diameters larger than the indicated particle diameter. For example, 84% of samples (by weight) have diameters greater than d_{84}, 16% have diameters less than or equal to d_{84}.

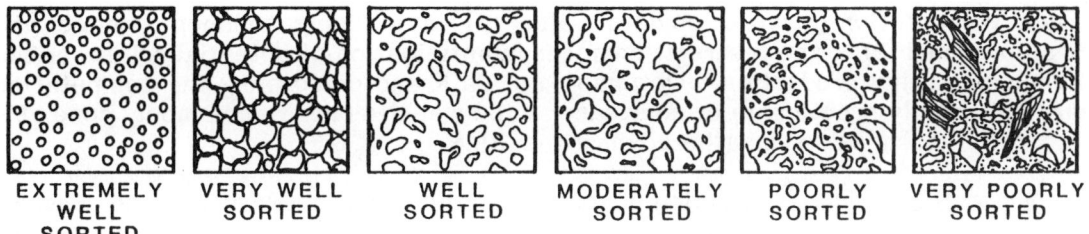

Figure 2-3. Comparative Sorting

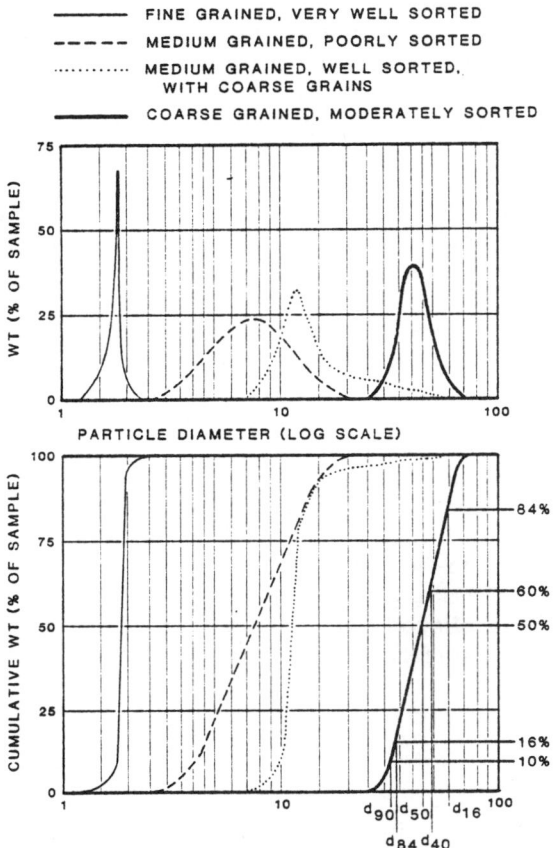

Figure 2-4. Mechanical Analysis for Size Distribution

Such computations are of obvious value (1) in studying sediment source and transport and (2) to the reservoir engineer in anticipation of potential solids control problems during production. Obtaining the data and performing the computation are not usually practicable or necessary at the wellsite but are commonly done at a laboratory using samples gathered during drilling.

In performing a visual evaluation of sorting, at the wellsite you should also consider grain shape. Extensive transport of sediment from its source with consequent erosion, abrasion and breakage is indicated by decreased grain size and increased sorting, roundness and sphericity. However, these factors are not always linked. A complete evaluation of all factors gives evidence of formation history and contributes to a complete qualitative estimate of formation porosity and permeability.

2.10 FABRIC

The arrangement of (nonspherical) particles within the rock gives it "fabric." In some rocks a preferred orientation may be observed, i.e., an imbricate texture indicating current strength and direction. In arenaceous rocks (particle size less than 2 mm) the lack of consolidation, or the small number of particles present in a single cutting, may prevent any recognition of the rock fabric. When examining core samples, include a description of rock fabric.

Although particle alignment cannot be seen in argillaceous rocks, it may be reflected in preferred break or fracture, linear color banding or in the luster of secondary minerals.

The gradation from mud or clay to shale is marked by a development of structure or fabric from soft, amorphous lumps to strongly fissile or foliated cuttings. Useful descriptive terms include:

- Amorphous
- Lumpy
- Massive
- Crumbly
- Banded
- Blocky
- Platy
- Laminated
- Jointed
- Fractured
- Flaky
- Hackly
- Splintery
- Sub-fissile
- Fissile
- Foliated
- Paper-Shale

2.11 STRUCTURE

The most important structure in rudaceous and arenaceous rocks is the bedding. This again may be indiscernible from cuttings in a single sample. Examining and comparing several samples through a section may allow bedding to be recognized. Cores often give a sufficiently large sample and vertical extent to allow deductions to be made as to the type and quality of bedding.

Bedding is divided into five major types for ease of description (Figure 2-5). These are not mutually exclusive, and a single bed may exhibit more than one characteristic:

- Regular or massive bedding
- Laminated bedding
- Graded bedding
- Current or cross-bedding
- Slump, or convolute bedding

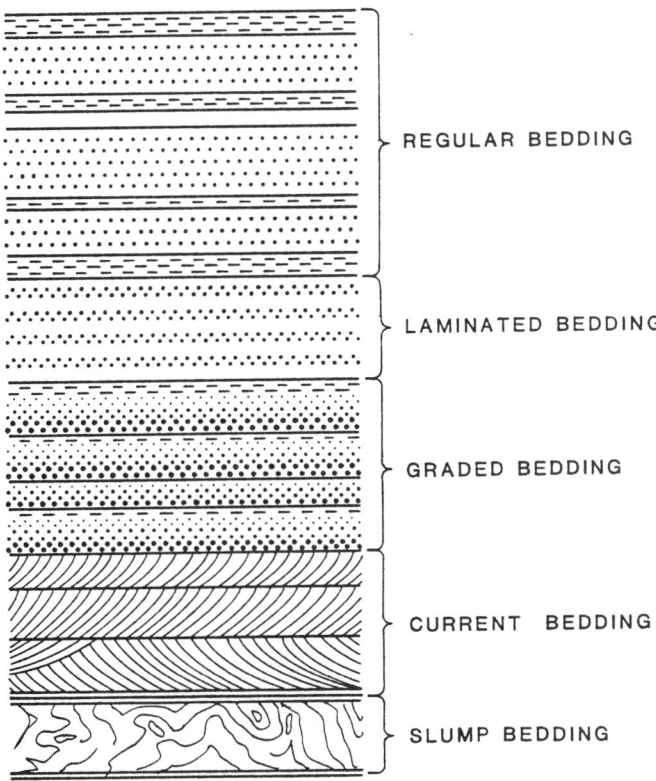

Figure 2-5. Types of Bedding

Regular bedding is indicated by parallel bedding surfaces, that is, those divisions in the lithology indicating a pause in the normal process of sedimentation. Between the bedding surfaces formations are normally uniform, indicating a constant source and transport of sediment. Beds on either side of bedding surfaces may be similar but rarely identical in all characteristics. The dip of the bedding surfaces is characteristic and structurally significant when recognized in a core in the same way as it would be in an outcrop. Remember that the long axis of the core represents a section of the formation cut parallel to the borehole and probably will not be vertical. Since rotation of the drillstring prevents any orientation of the core in a horizontal plane, the actual dip of the formation will be the dip relative to the core with a possible error of plus-or-minus the borehole inclination. Similarly, the strike of the bedding plane cannot be determined when core orientation is unknown, which is usually the case unless an oriented core barrel is used.

When examining the core, take care to discriminate actual bedding planes from breaks in the core (often in a vertical or subvertical sense) resulting from stress relief. A bedding plane indicates a halt in sedimentation. In addition to some change in the sediment, the plane itself may be marked by signs of drying or hardening, cracks, ripple marks, surface disturbance, detritus, or tracks. This halt may be of short duration and otherwise insignificant. Alternatively, it may be marked by signs of exposure or reworking prior to sedimentation beginning again.

Within regular bedding, laminations may be visible parallel to the bedding planes. Laminated bedding reflects minor variations in the nature of the sediment or the rate of sedimentation.

Graded bedding, while similar to laminated bedding, shows a systematic variation in grain size from coarsest at the base to finest at the top. Such gradation seen in a core or observed in a set of closely spaced cuttings samples will be a useful "way-up" indicator. Some minor mineralogical changes may also be present, but the smooth gradation in grain size is definite evidence of the continuous nature of the sedimentation and that no break has occurred. Graded bedding may result from a gradual consistent change in sediment source or current strength. Such consistency is unlikely, and graded bedding resulting from a gradually slackening current tends to be poorly sorted.

Graded bedding with well-sorted gradation is commonly associated with the settling in deep water of the vast quantities of sediment produced by a turbidity current. The settling of the mixed sediment load in effectively still water produces excellent sorting and smooth grain-size gradation. Graded beds of this type often occur in multiples, each graded bed being separated by a thin bed of a typical deep-water sediment (e.g., clays and claystones). Sole markings resulting from the movement of the turbidity currents may be preserved on the bottom of the graded bed or on the upper surface of the underlying bed, and these may be seen in cores. Further evidence of turbidity currents from unstable slopes may be the presence of fossils of shallow-water fauna in the graded bed, unlike the abyssal fauna seen in the interbedded deep-water sediments.

Current bedding may be present in aeolian, deltaic or other sediments and may commonly be recognized only in cores. Discrimination can be made by observing the grain shape and surface texture (aeolian grains tends to be better rounded, spherical and frosted) and by the sorting (air, having a lower competence than water, tends to produce finer, better sorted sediments). Curved depositional planes will be seen within the bed, and it is important to remember that the laminae defined by these will not be parallel to the bedding planes of the formation.

Usually, due to the impossibility of orienting a conventional core, it is not possible to determine the paleocurrent direction from current bedding as it would be in an exposure. The characteristic shape of the depositional planes, asymptotic to the erosion plane below and truncated by the erosion plane above, allows "way-up" identification.

Slump bedding or slump structure commonly results from the sliding and dislocation of a large bulk of sediment on an unstable slope. Layers of different sediments will become interfolded, broken up and mixed. Slump bedding is often complex and commonly can be detected only in an exposure. Even a core sample may be too small to interpret slumping, although small-scale structure within the major slump pattern can be discernible in a core.

The thickness of bedding of rudaceous sediments is commonly large enough to be seen only in cores, i.e., the thickness of a bed will be larger than the typical cutting diameter. Bedding in arenaceous rocks will be on a smaller scale but will be extremely difficult to detect conclusively in cuttings. It may, however, be seen in sidewall cores, or inferred from evidence of bedding planes and gradation or alternations seen in subsequent samples. Bedding thickness may be described as follows:

- Platy: less than 10 mm (0.5 inch)
- Very thin bedded: 10 mm to 50 mm (0.5 to 2 inches)
- Thin bedded: 50 mm to 100 mm (2 to 4 inches)
- Medium bedded: 100 mm to 300 mm (4 to 12 inches)
- Thick bedded: 300 mm to 900 mm (12 to 36 inches)
- Massive: greater than 900 mm (>36 inches)

Bedding which is not evident from cuttings cannot be recorded on the log. If bedding thickness is observable from later evidence (such as microresistivity logs), the information can be incorporated into a Final Well Report.

Strong bedding may sometimes be reflected in cyclical variations in rate of penetration or drilling torque over short intervals. Such evidence is of course not definite — it only indicates strong lineation in the rock. However, it may be useful in showing the true nature of a formation which becomes hydrated and unconsolidated in the drilling mud.

When describing bedding and bedding planes seen in cores, some associated and useful descriptive terms include the following:

Alternating	Cross-bedded
Well bedded	Wavy bedding
Crinkled bedding	Convolute bedding
Contorted bedding	Boudinage
Flame structure	Bioturbation
Lineated	Casts, moulds
Symmetrical, asymmetrical	

2.12 MINERALOGY

The mineralogy of a detrital sedimentary rock is a function of the parent rock and the means and severity of weathering, transport, and diagenesis.

2.13 Clastic

Rudaceous rocks are clastic in the specific sense; that is, they consist of fragments of previously existing rocks. It is important that the rock type or mineral from which the particles are formed be identified and described. This is important in the characterization of the rock itself and in identification of the source and possibly direction of transport of the sediment.

Fragments of igneous, metamorphic or sedimentary rocks may be present in a rudaceous rock, and all particles may represent a single parent rock source or may be a mixture of several. Limestone and coarse-grained igneous and metamorphic rock fragments will be chemically unstable in transport and burial, being susceptible to solution and various forms of chemical weathering and degradation. These reactions may result in changes in the particles themselves and in the matrix or cement of the rock. If possible, identify and describe the original form and weathered products.

Although diagenesis after burial modifies both the mineralogy and granularity of a rudaceous rock, two individual types (described in paragraphs 2.14 and 2.15) are readily recognizable.

2.14 Oligomict: These conglomerates are stable deposits typical of regressive shorelines. They consist of well sorted and rounded pebbles of a single rock type (commonly quartzite or chert), physically and chemically resistant to further weathering.

2.15 Polymict: These conglomerates and breccias show evidence of rapid deposition and burial close to the sediment source. Subangular and angular particles, poorly sorted both in terms of size and mineralogy, are contained in a matrix of similarly angular fine-grained particles or chemically derived residue. Having undergone little weathering in transport, the sediment is unstable and contains rocks such as limestones or igneous types which undergo numerous physical and chemical changes after burial.

When a rudaceous rock is close to its parent rock in terms of distance and mineralogy, discrimination between the two from cuttings samples may be difficult. Indeed there are cases where drilling may proceed from polymict breccia (e.g., a granite wash) via fractured and weathered parent rock, to fresh parent rock. In this circumstance, give greater attention to inspecting the sample for traces of weathering products, minor cementation and secondary mineralization, texture and staining on preexisting grain boundaries or fracture surfaces. Drilling parameters such as sharp fluctuations of rate of penetration and torque may also be of assistance in detecting the inhomogeneity of the breccia or scree. Similarly, hydrocarbon gases may be present in the secondary porosity which would not be expected in the unfractured parent.

2.16 Resistate

Arenaceous rocks may be clastic but in general are resistate, consisting predominately of quartz with minor feldspar and other detrital accessories. Little useful information can be obtained about the quartz mineralogy at the wellsite although, as stated above (paragraphs 2.6 through 2.9), the physical condition of the grains will yield information. The type, condition and abundance of minerals other than quartz will be of assistance in interpreting the environment and rate of sedimentation (paragraph 2.3) and may assist in isolating the source and history of the sediment. It will also serve the pragmatic purpose of improving identification of the sediment for later correlation.

Identification of rock mineralogy may also be important in selecting matrix properties for the interpretation of porosity and other wireline logs.

2.17 Hydrolysate

Argillaceous rocks and much of the matrix and secondary mineralization in rudaceous and arenaceous rocks are of hydrolysate origin, e.g. clay minerals, hydrous micas, hydroxides and some oxides. It is important to realize the subtle though significant difference between hydrolysate sediments and the other so-called "chemical" sediments. Hydrolysate minerals result from the chemical weathering of the parent minerals at the point of weathering and throughout the period of transport and sedimentation. True chemical sediments are produced by crystallization or precipitation at the place of sedimentation and may show no direct relationship to the parent, or parents, or the means of weathering and transport.

The five most significant minerals present in argillaceous rocks are the sheet silicates: illite, montmorillonite, vermiculite, kaolinite (all clay minerals) and chlorite. (Note: each of these mineral names encompasses a range of varying composition, i.e. a group of minerals related by a common structure. The term "smectite" has been used to describe the montmorillonite group, sometimes to include vermiculite. In keeping with the common literature of the subject this manual does not use the term, and use of a mineral name implies the mineral group.)

Clay minerals are usually the products of weathering and hydrothermal alteration of parent rocks, the latter probably being of lesser and possibly not quantitative importance (Grim, 1958). Acidic rocks poor in calcium, magnesium and sodium tend to yield kaolinite, whereas alkaline rocks generally yield montmorillonite. Illite may result from either rock type when potassium and aluminum concentrations are high. Chlorite is often detrital in sediments but may form from the degradation of ferromagnesian minerals. Vermiculite may result from the degradation of micas and is also present in a mixed-layered form with detrital or secondary chlorite.

In addition to the sheet silicates, fractions of accessories include unaltered parent minerals and resistant material, e.g. quartz. Reworked, previously compacted and reweathered clay minerals may also be present. The presence or absence of these in quantity is symptomatic of the energy and activity of the environments of weathering, transport and sedimentation. Since the physico-chemical weathering process is continuous, conditions within the environments of weathering, transport and sedimentation have as large if not a larger effect on the mineral product as the parent.

Weaver (1958) generalized certain environmental parameters for clay mineralogy (Figure 2-6). These generalizations, in practice, prove insufficiently refined for isolated application. Weaver (1960) states that "...when clay petrology is integrated with other available geological data, the cause usually becomes apparent." Even in ancient clay rocks, despite diagenesis, sediment source and history may be recognizable in clay mineralogy. Thus once the clay-environmental parameters have been recognized for a region, clay mineralogy may be used independently or in conjunction with other evidence to identify changes in environment or source resulting from climatic, topographic or orogenic events.

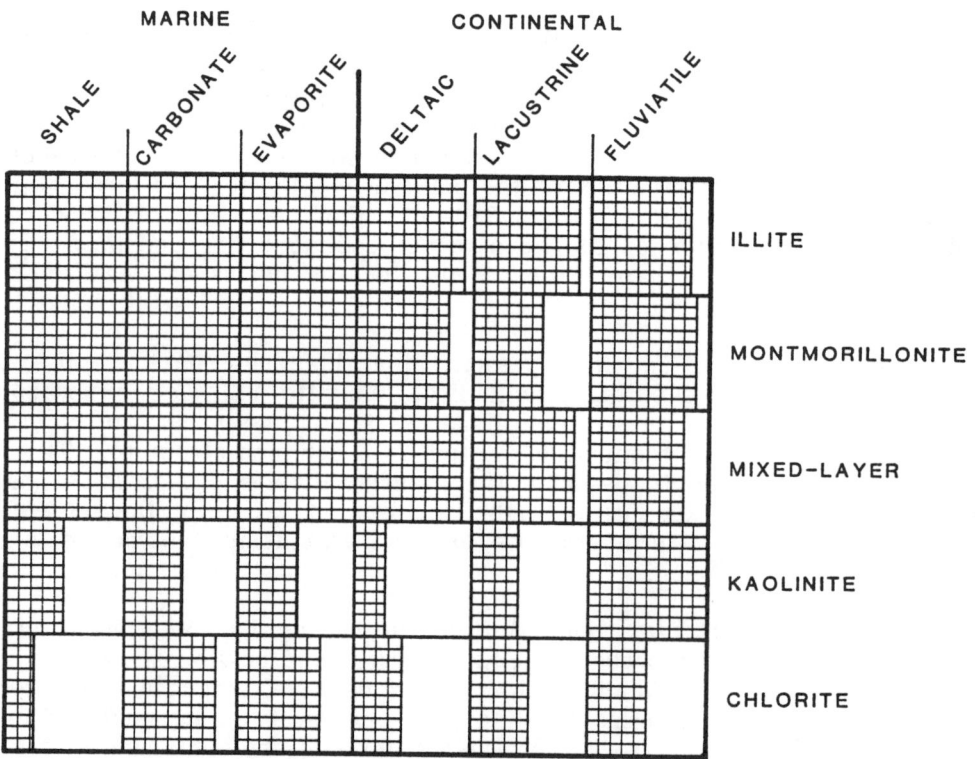

Figure 2-6. Types and Amounts of Clay Minerals Commonly Found in Sediments

With drill cuttings samples from the Norwegian North Sea, Karlsson et al (1978) used changes in the clay mineral suite, based upon bulk chemistry and cation exchange capacity, to monitor changes in climate, sediment source and erosion rate in Tertiary argillaceous rocks (Figure 2-7).

Determination of Cation Exchange Capacity (C.E.C.) of drilling muds has long been a common tool for estimating the bentonite content. The test involves titrating diluted drilling mud with methylene blue. Application of this method for monitoring changes in clay content and mineralogy is available from Exlog as a secondary service (commonly referred to as the "Shale Factor Test"). The principle of the test is that the volume of methylene blue required to reach an end-point (milli-equivalents of methylene blue per 100 grams of sample) is a function of the number of free ion exchange sites available within and on the clay lattice. Since these sites are fixed by potassium ions during illitization, the C.E.C. measurement is a measure of the relative montmorillonite and illite composition in a young

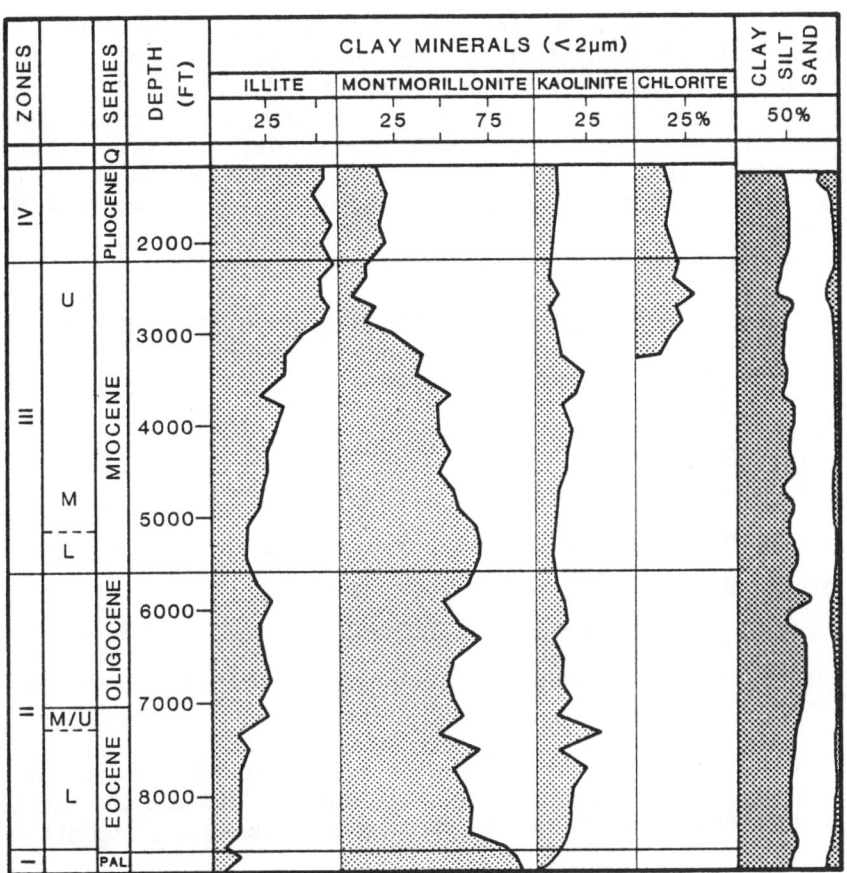

Figure 2-7. Mineralogical Variation in the Tertiary Section of Norwegian Well 2/11-1 (after Karlsson et al, 1978)

sediment and a monitor of the process of diagenesis in older rocks. Arguments raised against its use include the inability to monitor changes in the nonclay fraction of the clay rock, drilling fluid and caving contamination, and variable reaction rate with solid clay particles of varying size and permeability.

These are indeed problems to consider when interpreting data from the C.E.C. test. On the other hand, it should be pointed out that the test is producing quantitative results, indicating changes in a lithology which, from all other cuttings observations, would remain uniform and lacking in distinction. Careful sample selection, grinding and dispersion can remove most of the sampling and titration errors. Where calcimetry (acidometric determination of calcite and dolomite) is being run at the wellsite, a correction for carbonate content may be made. Alternatively, acidizing and sieving a dispersed clay sample assist in removing carbonate and clastic material prior to performing the test. Comparison of sidewall core and cuttings results (Karlsson et al, 1978) confirms theoretical predictions that exposure of the cuttings to the drilling mudstream produces no significant modification to clay mineralogy and test results.

The difference in C.E.C. of the clay minerals (illite, 30 meq/100 g; montmorillonite, 100 meq/100 g) is sufficient for the test to be very sensitive to changes in clay mineralogy. Karlsson et al (1978), using C.E.C. with additional chemical analyses, converted the results to relative composition of the clay minerals (see Figure 2-7). This treatment, while helpful, is not necessary and does involve additional measurements and calculations which may not be possible or desirable to perform at the wellsite. Figure 2-8 readily demonstrates that the distinct mineralogical zonation shown in Figure 2-7 is equally if not more apparent in the C.E.C. data alone.

If a Shale Factor Test Kit is not available at the wellsite, a simple "wettability" test will provide diagnostic and correlatable information. Clays may be characterized by placing a single cutting in a cut dish, adding distilled water and inspecting the reaction through the microscope. The responses may be described as follows:

- Hygroturgid: swelling in a random manner
- Hygroclastic: dispersing as irregular fragments
- Hygrofissile: separating into tabular flakes
- Nonswelling: no reaction
- Cryptofissile: separating into tabular flakes when 10% hydrochloric acid is added

If the clay is calcitic, swelling after addition of acid may be due to solution of calcite rather than to a reaction of the clay minerals. Inspect the debris after the reaction is complete to determine whether the rock was clay or calcite supported.

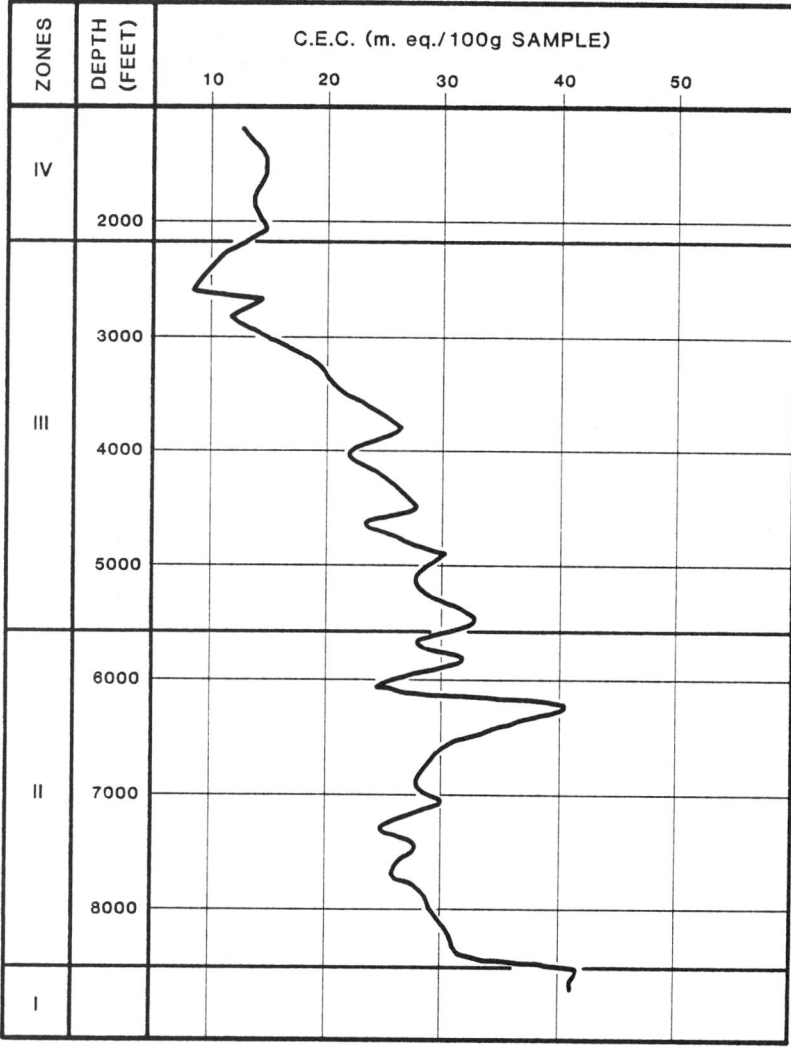

Figure 2-8. C.E.C. Plot for the Tertiary of Norwegian Well 2/11-1
(Data from Karlsson et al, 1978)

2.18 INDURATION

Induration or lithification is the process by which a sediment is converted into a sedimentary rock and is the result of all those processes encompassed by the term diagenesis. In detrital rocks, diagenesis and induration consist mainly of the crystallization of minerals chemically derived from the particles, the matrix, or externally to form a cement. Since the particles themselves are of a relatively uniform hardness, induration will be a function of the type and quantity of this cement. Typical descriptive terms include:

Brittle	Blocky	Plastic
Hard	Splintery	Soft
Firm	Earthy	Unconsolidated
Dense	Friable	Amorphous
Crumbly	Loose	Soluble

Remember that the term is descriptive of the rock as a whole. Thus quartz is (relatively) hard in whatever form it may occur, and yet a quartz-cemented quartz sandstone may range from an unconsolidated sand (will colapse under its own weight) to a britle orthoquartzite (will fracture across grain boundaries), depending upon the quantity, crystallinity and distribution of the cement.

Like sorting, rock "hardness" is a distinct measurable quantity but one which under normal circumstances cannot be measured under wellsite conditions and time constraint. A useful qualitative guide to rock induration is given below. (Mineral hardness is a distinctly different quantity and is discussed in paragraph 2.25.)

2.19 Clastic and Resistate

- Unconsolidated: Cuttings fall apart, or sample occurs as individual grains
- Friable: Rock crumbles with light pressure. Grains easily detach with sample probe
- Moderately Hard: Grains detach with sample probe. Cuttings can be broken with some pressure
- Hard: Grains difficult to detach. Extreme pressure causes cutting to break between grains
- Extremely Hard: Grains cannot be detached. Cuttings will break through grains

2.20 Hydrolysate

- Soluble: Readily dispersed by running water
- Soft: No shape or strength. Material tends to flow
- Plastic: Easily molded and holds shape. Difficult to wash through a sieve
- Firm (frangible): Has definite shape and structure. Readily penetrated and broken by probe
- Hard: Sharp angular edges. Not readily broken by probe

2.21 CEMENT

The character and composition of the cementing material are of prime importance in anticipating reservoir performance. They may also indicate depositional history, predepositional weathering and postdepositional alteration. Cementation may be derived from or related to the rock particles or may be externally derived.

Common cementing materials are:

- Calcite
- Sulfates
- Silica
- Clays
- Siderite
- Dolomite
- Iron Oxides
- Pyrite

Although induration is related to the quantity of cement, the quality of cement is also a factor. For example, in a quartzose rock, pressure solution and recrystallization at grain boundaries may lead to a rock becoming indurated and hard without large amounts of cement being discernible.

The difference between "cement" and "matrix" is one of degree and may not readily be obvious in any one sample. Gradations may occur. In general it can be said that, where intergranular contact does not occur, the fill material between grains is a matrix. It cannot be denied, however, that this matrix does have a cementing function in holding the grains fixed relative to each other. This is true even in the case of clays. Figure 2-9 shows a gradation between a sandy mudstone and an orthoquartzite and points out the ambiguity between matrix and cement.

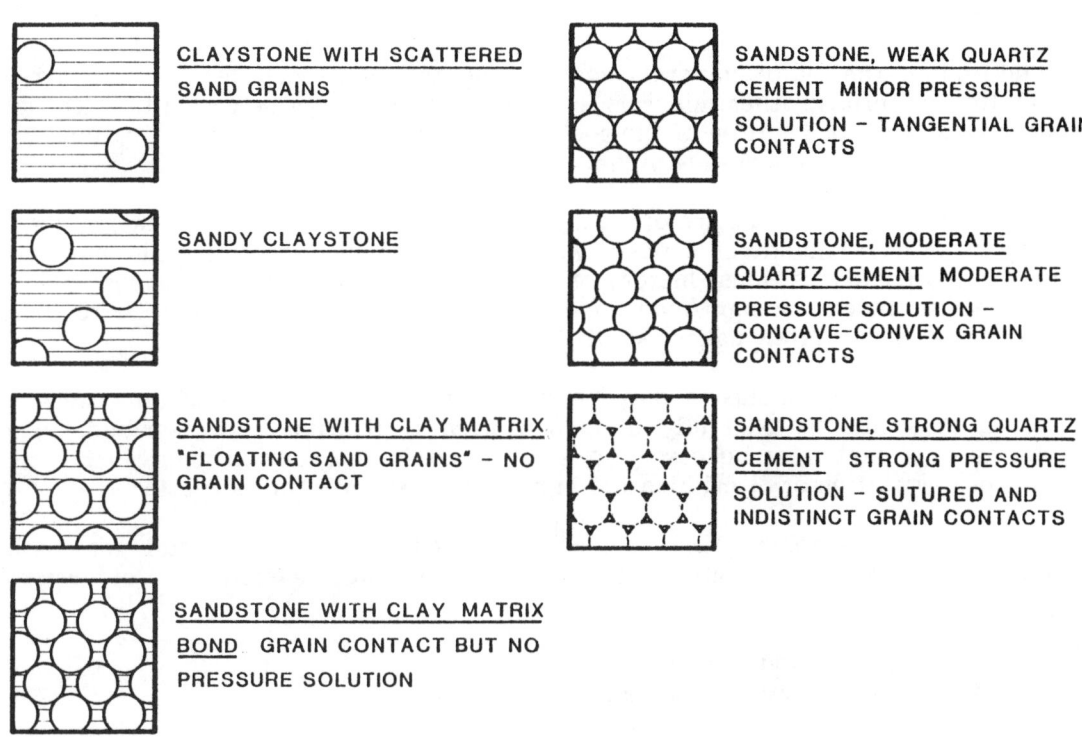

Figure 2-9. Cement and Matrix

An alternative method of discrimination is to define matrix as primary sedimentary material, whereas cement is secondary, chemically precipitated material. This definition may be difficult to apply when primary sedimentary material such as clays or calcite shows evidence of recrystallization. When difficulty occurs it is best to describe the material as cement/matrix.

2.22 MATRIX LITHOLOGY

In addition to cement and accessory minerals, rudaceous and arenaceous rocks may contain large amounts of sedimentary material between the rock particles. This is especially true of the more coarse-grained type. The material may consist of fill within the rock voids or may be sufficient to actually separate the coarse particles. Indeed, it is common for a bed to grade from a psephitic rock with a fine-grained matrix to rock consisting solely of that fine-grained sediment. In such cases it is necessary to treat the matrix as a separate lithology and describe it in as much detail as if present alone, e.g., Pebble Conglomerate, etc., in a matrix of Sandstone, etc.

2.23 ACCESSORIES

2.24 Minerals

In addition to the major minerals constituting the rock particles and the major cementing materials, other minerals in trace quantities may be present. Although constituting a minor fraction of the rock bulk, accessories are of disproportionately great diagnostic and descriptive value.

Detrital accessory minerals are often diagnostic of the source of sediment, mode of transport or environment of deposition. Diagenetic accessory minerals may give clues to the postdepositional history of the rock. Even if such conclusions are not immediately obvious or significant to the logging geologist, they may be to some other geologist who later makes use of the log.

Though no specific conclusion may be drawn from the presence of an accessory mineral, it should be carefully described even if it cannot be ascribed. The presence of a particular accessory at some point in the section may be the only means of correlation or discrimination in an otherwise uniform succession.

Make a note of secondary enlargement of detrital grains or crystal growth in diagenetic minerals. Common terminology for the development of crystal structure is:

- Anhedral -- no visible crystal form
- Subhedral -- partly developed crystal form
- Euhedral -- well developed crystal form

Common accessory minerals in sedimentary rocks are illustrated in Figure 2-10. Identification of a particular mineral may be difficult in the fine-grained, weathered state common in a sediment. Even if a mineral cannot be conclusively identified, it should be sufficiently described to allow later recognition.

	MINERAL	DENSITY (S.G.)	HARDNESS (MOH'S)	OBSERVABLE FEATURES	OCCURRENCE
VERY ABUNDANT	Orthoclase	2.56 to 2.63	6 to 6.5	Colorless or white, sometimes pinkish-yellow (flesh colored), insoluble in acid	Detrital from acid and basic plutonic rocks
	Plagioclase	2.62 to 2.76	6 to 6.5	Gray, rarely red; albite twinning on fresh cleavage faces; partially soluble in acid	1. Detrital from basic and intermediate extrusive rocks and high grade metamorphics 2. Pressure solution of detrital grains and re-crystallization in cavities and fossil replacement
	Quartz	2.65	7	White, sometimes smoky gray or colored by inclusions, rarely black; glassy luster	1. Detrital from all rock types 2. Secondary crystallization around detrital nuclei
	Opal/Chalcedony	2.55 to 2.61	7	White, milky; spherulitic	Deposition from colloidal silica
ABUNDANT	Biotite	2.7 to 3.3	2.5 to 3	Translucent, black, brown-red/brown; high luster; strong cleavage; weak-moderate magnetic	Detrital from acid plutonic and metamorphic rocks
	Calcite	2.71	3	Colorless-white; vitreous luster; rhombohedral cleavage; soluble in cold acid	1. Recrystallization of organic remains 2. Chemical precipitate 3. Detrital from basic igneous rocks
	Kaolinite	2.61 to 2.68	2 to 2.5	White; flakes or plates formed into compact, friable, "mealy" masses; greasy, texture; insoluble in acid	Hydrothermal weathering of acid igneous rocks
VERY COMMON	Chlorite	2.6 to 3.3	2 to 3	Green; flaky mica-like appearance; decomposed by sulfuric acid	1. Low grade metamorphosis 2. Hydrothermal alteration of igneous rocks 3. Maturation of clays
	Garnet:		6 to 7.5		
	Pyrope	3.582		Deep Red-black	Detrital from ultrabasic ignous rocks
	Almandine	4.318		Deep red-red/brown	Detrital from medium to high grade metamorphosed argillaceous rocks
	Spessartine	4.190		Black-red/brown-orange	Detrital from granitic and associated metasomatic rocks
	Grossular	3.594		Pale green-yellow; sometimes white	Detrital from metamorphosed impure calcareous rocks
	Andradite	3.859		Golden yellow-black	Detrital from metamorphosed impure calcareous and calcic igneous rocks
	Uvarovite	3.900		Dark green	Detrital from Serpentines

Figure 2-10. Accessories Occurring in Sedimentary Rocks

	MINERAL	DENSITY (S.G.)	HARDNESS (MOH'S)	OBSERVABLE FEATURES	OCCURRENCE
VERY COMMON	Hydrogrossular	3.13 to 3.594		Red/brown; dodecahedral crystal form or as spherical masses or grains; weakly magnetic	Detrital from metamorphosed marls and altered gabbroic rocks
	Hornblende	3.02 to 3.45	5 to 6	Dark green-black; good cleavage; weakly to moderately magnetic	Detrital from all igneous and metamorphic rocks
	Ilmenite	4.70 to 4.78	5 to 6	Black; rarely with red/brown tinge; sub-metallic luster; embedded masses or irregular-hexagonal plates; difficultly soluble in acid; moderately magnetic; may be distinguished from magnetite by presence of greyish white alteration product, Leucoxene	Detrital from many igneous and metamorphic rocks
	Limonite	2.7 to 4.3	4 to 5.5	Yellow/brown-dark orange/brown; earthy; occasionally vitreous; "varnish-like" coating; slowly soluble in hydrochloric acid; yellow streak	1. Alteration product of iron-bearing minerals 2. Biogenic deposit
	Magnetite	5.20	6	Black-dark gray; opaque; brittle; fine-dull metallic luster; small grains lacking structure; strongly magnetic	1. Detrital from many igneous rocks 2. Thermally altered sediments
	Muscovite	2.77 to 2.88	2.5 to 3	Colorless-light brown/green; high luster; strong cleavage; may be difficult to distinguish from Biotite if color is not discernible	1. Detrital from acid igneous and associated metamorphic rocks 2. Low grade phyllites and schists
	Pyrite	4.95 to 5.03	6 to 6.5	Brassy yellow; occasionally black metallic luster; conchoidal-uneven fracture; cubic or pyritohedral crystal form	1. Hydrothermal veins 2. Detrital from metasomatic phyllites 3. Biogenic and diagenetic in muds
	Zircon	4.6 to 4.7	7.5	Red/brown, yellow-grey/green; tetragonal crystal form	Detrital from sodium rich plutonic rocks. May survive several cycles of weathering and deposition
COMMON	Actinolite	3.02 to 3.44	5 to 6	Gray-bright green; opaque-translucent; vitreous luster; may occur as fine fibrous growths	Detrital from contact and regional metamorphic rocks
	Andalusite	3.13 to 3.16	6.5 to 7.5	Pink; may be white-rose/red; subtranslucent; brittle, splintery	Detrital from metamorphosed argillites
	Augite	2.96 to 3.52	5 to 6	Dull green-brown/black; presence of opaque black iron weathering products will distinguish from hornblende	1. Detrital from gabbros, dolerites, and basalts 2. Detrital from metamorphosed Limestones
	Cassiterite	3.98 to 4.02	9	Red/brown-black; adamantine luster; slowly dissolved by acids	Detrital from tin-bearing acid igneous rocks
	Chromite	5.09	7.5 to 8	Red, brown, black, green; high luster; pithy, rarely of megascopic size	Detrital from basaltic and ultramatic igneous rocks
	Corundum	3.98 to 4.02	9	Dark blue/gray; smoky; adamantine-vitreous luster; translucent-opaque; grains or shapeless lumps	Detrital from alkaline and silica-poor metamorphic rocks

	MINERAL	DENSITY (S.G.)	HARDNESS (MOH'S)	OBSERVABLE FEATURES	OCCURRENCE
COMMON	Enstatite	3.21 to 3.96	5 to 6	Grey or green, yellow-brown; similar to Augite but iron-poor	1. Detrital from ultra-basic igneous rocks 2. Detrital from medium grade metamorphosed argillites
	Epidote	3.38 to 3.49	6	Olive-yellow/green; opaque-translucent; vitreous luster; bundles of bladed prisms or needles, slow reaction with acid	Detrital from metamorphosed basic igneous rocks
	Glaucophane	3.08 to 3.30	6	Lavender-deep blue; similar to Hornblende; distinguished by color	Detrital from highly deformed meta-sediments, e.g., greenschists, metagrey-wackes
	Gypsum	2.30 to 2.37	2	White or colorless; occasionally with red or blue tinge; white precipitate with barium chloride; distinguished by density and hardness.	1. Dehydration of sea water 2. Groundwater alteration of calcium carbonate
	Anhydrite	2.90 to 3.0	3 to 3.5		
	Kyanite	3.53 to 3.65	5.5 to 7	White-pure blue; vitreous or pearly lustre; bladed crystals or columnar masses	Detrital from metamorphosed sandstones
	Monazite	5.0 to 5.3	5	Yellow-red/brown; spherical masses or grains	1. Detrital from granitic rocks 2. Detrital from dolomitic marble
	Rutile	4.23 to 5.5	6 to 6.5	Red/brown; may be black, violet, green; fine needle-like crystals in shale	1. Detrital from granite pegmatites and quartz veins 2. Detrital from metamorphose argillites 3. Maturation of clays and shales
	Staurolite	3.74 to 3.83	7.5	Blood red-yellowish brown; stout thick crystal; commonly associated with garnets	Detrital from medium grade metamorphosed argillites, grits and carbonates
	Titanite	3.45 to 3.55	5	Colorless, yellow, green, brown; rhombic cross section	1. Detrital from intermediate and acid plutonic rocks 2. Detrital from impure calc-silicate metamorphic rocks 3. Possibly (?) diagenetic in sandstones
	Topaz	3.49 to 3.57	8	Colorless; rarely yellow-brown or white; brittle with uneven fracture	1. Detrital from acid ignous rocks 2. Detrital from metamorphosed bauxite
	Tourmaline	3.03 to 3.25	7	Black; very rarely green, brown, red; opaque; glassy-dull luster, long, thin prisms, with curved triangular cross section	1. Detrital from granitic rocks 2. Detrital from metasomatised basic igneous rocks 3. Secondary mineral growth on detrital grains in sandstones 4. Replacement in Limestones

	MINERAL	DENSITY (S.G.)	HARDNESS (MOH'S)	OBSERVABLE FEATURES	OCCURRENCE
RARE	Apatite	3.1 to 3.35	5	White, green, brown; transparent-translucent; vitreous luster; crypto-crystalline concretionary masses in phosphatic beds	1. Detrital from all igneous and metamorphic rocks 2. Diagenesis of phosphatic deposits, e.g., fish scales, bones
	Aragonite	2.94 to 2.95	3.5 to 4	Colorless-white; reacts with acid but denser than calcite	1. Organic remains 2. Chemical precipitation 3. Altered Dolomite
	Barite	4.5	2.5 to 3.5	White-yellow/gray, blue, pale green, vitreous-resinous luster, globular concretions or fibrous, earthy aggregates; sometimes well-formed tabular crystals (of desert roses), evolves Hydrogen Sulfide with gentle heating (german name "Stinkstein"), and colors flame blue	1. Hydrothermal veins 2. Cavity growth in Limestone 3. Limestone weathering product 4. Hot Springs
	Fluorite	3.18	4	Colorless, white, yellow, green, blue, purple; vitreous luster; similar to calcite but poorly reactive with Hydrochloric acid, soluble in Sulfuric acid	1. Detrital from acid igneous rocks 2. Hydrothermal deposit 3. Deposition by circulating mineral rich waters
	Galena	7.5 to 7.6	2.5	Dull gray; metallic luster; distinguished by high density and softness	1. Hydrothermal veins 2. Low temperature hydrothermal deposits in sediments
	Glauconite	2.4 to 2.95	2	Colorless-yellow/green-blue/green; similar to biotite but forms rounded aggregates of plates	Marine diagenesis in shallow slow-depositing basins
	Hematite: Specular Iron Ore Micaceous Hematite Common Red Hematite	5.256	5 to 6	Black steel gray; rarely reddish; opaque; bright metallic luster; masses and sub-hexagonal plates; subconchoidal fracture Deep red; translucent-opaque submetallic luster; mica-like flakes with ragged sub-hexagonal outlines Dark red; dull luster; cryptocrystalline columnar or granular masses	1. Weathering product of iron bearing minerals 2. Metamorphism of iron-rich sediments
	Sphalerite	4.1	3.5 to 4	Black, brown, yellow, red green; resinous or adamantine luster brittle; tabular or fibrous growths	1. Low temperature veins in dolomitic limestone 2. Chemical accretions in sandstones and limestones
	Vesuvianite Idocrase	3.33 to 3.43	6 to 7	Apple green, yellow, brown; vitreous luster; square prismatic crystal form; may occur as irregular lumps	1. Detrital from contact metamorphosed Limestone 2. Detrital from veins in basic and ultrabasic rocks
	Zoisite	3.15 to 3.27	6	Gray; aggregated blades or prismatic masses	1. Detrital from medium grade metamorphosed marls 2. Detrital from thermal metamorphism of impure limestones 3. Hydrothermal alteration of Plagioclase

2.25 Moh's Scale of Hardness

This is not a quantitative scale (e.g. Rockwell Hardness) but defines an order of hardness useful for testing minerals. A sharply pointed mineral fragment will scratch any other mineral having a lower number on the scale. Thus a mineral which scratches gypsum but is scratched by calcite has a hardness of 2-1/2.

Standards	Moh's Scale	Useful Tests
Talc	1	
Gypsum	2	
		<------------------------ Fingernail
Calcite	3	
		<------------------------ Brass pin
Fluorite	4	
Apatite	5	
		<------------------------ Pocketknife
		<------------------------ Window glass
Orthoclase	6	
Quartz	7	
Topaz	8	
Corundum	9	
Diamond	10	

2.26 Fossils

Other than detrital or secondary minerals, fossils form the other common "accessory" in sedimentary rocks. In clastic rocks whole macrofossils are rare, and in cuttings samples only broken fragments will be seen. Fossiliferous formations will probably yield only unidentifiable macrofossil fragments or microfossils too small to be identified under the magnification power available to the logging geologist. A note as to the presence of fossils, a brief description or a tentative identification, perhaps phylum only, will be useful to other geologists reviewing the log and may even be of assistance to a paleontologist who later examines the cuttings. The amount of fossil material in a sample should be estimated as follows:

>25%: Abundant
10 - 25%: Common
0 - 10%: Rare

Micropaleontology is too extensive a subject to be given adequate coverage. Logging Geologists are referred to their local office library to obtain further information.

2.27 COLOR

The color of coarse-grained rocks such as the rudites is of less diagnostic value than is normally the case in rock description. The grain size itself is indicative of sedimentary environment, near to shore and energetic. Furthermore, color of the rock is controlled by the color of the sediment source rock.

Pelitic sediments, being composed predominately of clay minerals, should (if pure) be white or pale gray. Color is given to the sediments by the presence of iron, sulfides or carbonaceous material. Color is therefore a useful environmental indicator in argillites; for example:

- Red, brown: ferric iron, an oxidizing environment
- Green, grey-blue: ferrous iron, a reducing environment
- Dark brown, black: organic material, a potential petroleum source
- Dull black: sulfides, an anaerobic environment conducive to the production of hydrocarbons.

For each sample give a detailed color description, distinguishing the colors of

- Particles
- Matrix or cement
- Staining or secondary mineralization
- Accessories

Even when of no diagnostic significance, color is a major criterion for recognition and hence correlation, so color descriptions should be graphic and complete. Use comparative terms if they can give a better representation of the color being described. For example,

- Chocolate brown
- Brick red
- Bottle green

When estimating color, inspect both wet and dry samples. The wet sample will reveal colors more vividly and should be the basis of a color assessment. Dried cutting may sometimes allow a better discrimination of subtle hues and color shades.

Color distribution is also significant, and suitable terms are

- Multicolored
- Banded
- Irridescent
- Scattered
- Speckled
- Spotted
- Disseminated
- Variegated

Color changes in a sediment may be very subtle. For this reason, wet samples should be retained long enough so that, at anytime, at least 100 feet of sample may be viewed together for comparison. When first coming on tour, review samples from the previously drilled 100 feet prior to looking at fresh samples.

2.28 PETROLEUM SIGNIFICANCE

Detrital rocks are important in petroleum accumulation as source rocks, reservoir rocks and cap rocks. Evaluation for oil and gas must be made for all samples. As time and drill rate allow, a complete lithological and hydrocarbon evaluation, as described in paragraphs 1.9 through 1.16, should be performed regularly.

In the past it was common to dismiss argillaceous sediment as uniform gray, green or brown massive zones of varying hardness but of little further economic value or geological interest. This attitude has now been overturned by the realization that such zones are complex geochemical systems providing most of the hydrocarbons and much of the water involved in the migration and accumulation of a petroleum reservoir.

Argillaceous sediments undergo many changes in burial beyond simple compaction and induration. These processes and their physical effects are discussed in <u>Theory and Evaluation of Formation Pressures: A Pressure Detection Reference Handbook</u> (EXLOG, 1985).

In paragraphs 2.29 through 2.31 the geochemistry of clay diagenesis is reviewed. Special attention is given to its broader geological significance and to the generation and migration of hydrocarbons.

2.29 Clay Diagenesis

There is some controversy as to the conversion processes which take place in a clay rock after burial. Indeed, the term "diagenesis" itself is claimed to be a misnomer for the processes which can and do occur under temperature, pressure and other conditions prevalent in a sedimentary basin — the transformation of kaolinite, montmorillonite and vermiculite clays to illite and chlorite (Figure 2-11).

The basic clay lattice structure has sufficient stability to remain unchanged except under chemical or physical environment extremes in weathering or metamorphism. On the other hand, the exchange cations adsorbed onto the clay lattice sheets, as a result of the unsatisfied charge, are only weakly bound and readily interchange with other ions in response to changes in chemical environment. Some cations are more strongly bound and may be preferentially retained, giving the appearance of a permanent chemical or mineralogical change.

It has been noted that mineralogical changes promoted in weathering may be reversed by environmental changes during transport and deposition (Weaver, 1958). Following burial, however, environmental changes are generally progressive and monotonic. Recovery of drill cuttings from an exploration borehole, using a well-designed, nonreactive drilling fluid, is rapid and lacking in chemical extremes. For these reasons, when studying mineralogical changes with depth of argillaceous rock cuttings, it is valid both in terms of geology and sampling technique to consider the conversion processes (shown by arrows in Figure 2-11) to be irreversible. That is, once a conversion occurred in the processes of burial and consolidation, it cannot have been reversed during the continuance of that process, nor by the process of drilling and recovery to surface. Thus the mineralogy seen in a cutting or core sample is indicative of the maximum degree of conversion completed in that formation. This assumption is supported by both theoretical and practical studies (Weaver and Beck, 1971; Karlsson et al, 1978). Adsorption and osmotic phenomena do occur, and water content may be drastically modified by drilling fluid reactions.

At deposition and after initial burial, a clay sediment contains approximately 20 percent by volume of mineral material. This consists predominately of randomly interstratified illite and montmorillonite. Very rarely does either mineral occur in

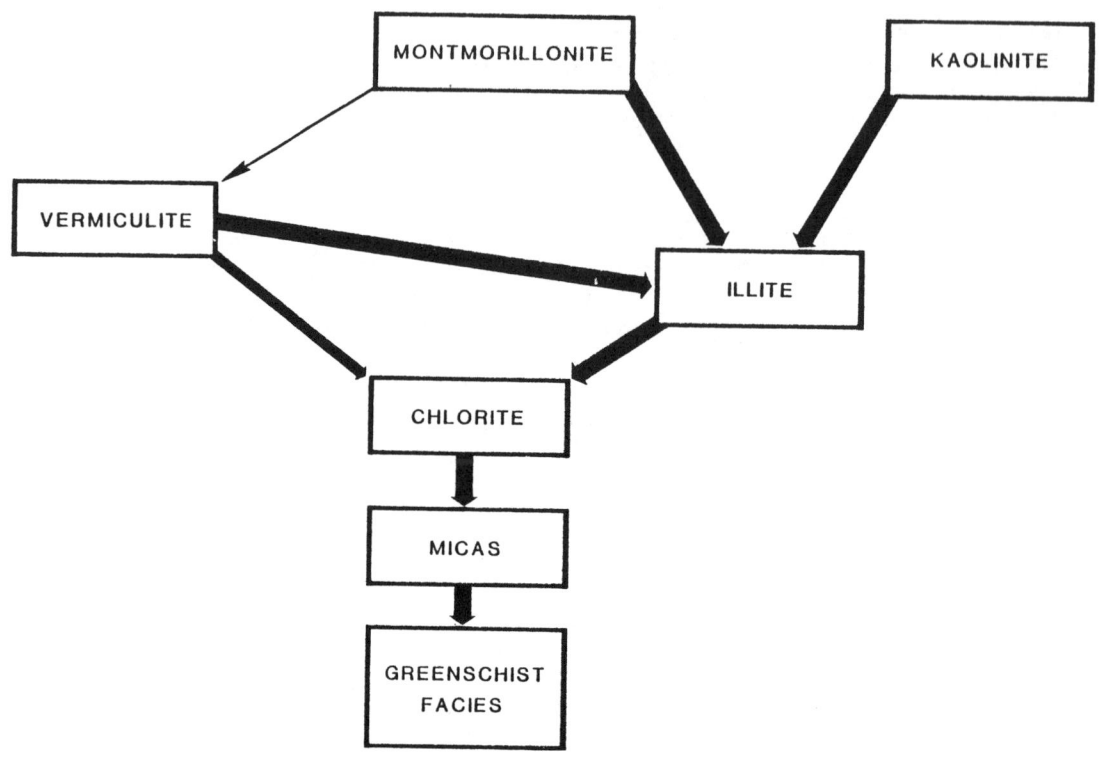

Figure 2-11. Diagensis and Metamorphism of Sheet Silicates

a pure form but usually as a randomly mixed-layered mineral with the proportion of illite and montmorillonite varying greatly between samples. Kaolinite, chlorite and vermiculite occur in smaller quantities depending upon the sediment source and history, and the remainder of the mineral phase consists of detrital minerals and hydrated weathering products, oxides, and amorphous inorganic and organic materials.

The approximately 80 volume percent of water in the sediment is not present in a continuous fluid phase but is present in three distinct forms:

1. Interstitial pore water between and, to a large extent, supporting and suspending the mineral particles. This water is in a free fluid form and has both suspended and dissolved ionic and organic content.

2. Interlayer water within the lattice space of expandable montmorillonite clay layers. Although chemically bound to the clay lattice, the water is relatively mobile and may be removed by compaction. Being chemically bound to the clay mineral, this water cannot be said to have a salinity; but a free interchange of ions, organic and water molecules takes play between the interlayer and interstitial water.

3. Water of hydration is more strongly bonded to both the expandable clay lattice and to the hydrated inorganic material. This water is immobile and cannot be removed by the application of pressure alone.

Of the environmental factors acting during burial, it appears that overburden pressure, temperature and ionic concentration are the most critical parameters governing diagenesis. Depth of burial is important only because of its relationship to temperature and pressure. Age is apparently of little or no significance.

During initial burial, only physical dehydration occurs without mineralogical changes. Under overburden loading, interstitial water and interlayer water in excess of the two layers constituting water of hydration are hydrodynamically "squeezed" from the sediment. This large water efflux, freshened by the admixture of interlayer water, dissolves and flushes a proportion of the amorphous inorganic material, but the less soluble and more particulate mineral and organic detritus are little affected. This is a continuous process with increasing overburden load.

During this stage most mobile interlayer water is removed, leaving about 20 volume percent water of hydration. Although increasing overburden with burial gives added impetus to dewatering, decreasing porosity produces a radically greater reduction in permeability. The overall result is that water loss rate and porosity decline with increasing depth, coming effectively to a stop with approximately 10 percent interstitial water.

True diagenesis begins only at greater depth as a function of rising temperature. The process involves the fixation of potassium ions at the lattice sites within montmorillonite clays. Upon fixation, the lattice spacing contracts or "collapses" to the spacing typical of illite, forcing the two layers of water or hydration out into the interstitial space. Experimental evidence suggests that the molecular arrangement of the water of hydration when hydrogen-bonded to the clay lattice is more compact than that of liquid water. Therefore, the required increase in interstitial volume to accommodate the water is greater than the reduction in matrix volume resulting from desorption; that is, to accommodate the extra water volume, porosity must in theory be created.

Creation of porosity on a large scale is neither theoretically justifiable nor observed in nature. Thus where the hydraulic conductivity of the rock is exceeded, abnormally high fluid pressures initiate and utilize new hydraulic flowpaths for water expulsion. It has been suggested that it is at this stage fissility and jointing are first developed in the rock.

The potassium ions required for illitization may be derived from the detrital and degraded solid phases intimately associated with the clay via the interstitial water, potash feldspars, micas, oxides and hydroxides. Therefore, diagenesis does not necessarily require an influx of water or ions into the rock. Indeed, the efflux of water desorbed during diagenesis carries with it large quantities of potassium, other exchange cations and silica which later constitute the cementing material found in aquifers and indurated rocks.

In parallel with this conversion, kaolinite and vermiculite conversion to illite and chlorite by the incorporation of detrital ionic aluminum and iron is also occurring. These processes do not, however, result in the desorption of such large quantities of water.

In summary, the conversion process is predominately temperature-dependent such that, in areas of lower geothermal gradient, diagenesis occurs at greater depth and over a greater depth range. However, adjustment of data for geothermal gradient changes does not produce equivalent results (Perry and Hower, 1972), indicating that depth of burial and hence pressure does have an influence on the process. Since the reaction involves a net volume increase, it appears likely that where sufficient dewatering paths are available, pressure acts to accelerate reaction; but where they are not, or must be created, it seems pressure would have a retarding influence. Finally, the availability of mobile potassium ions, as a reaction component, radically affects reaction rate. In fact, the presence of high potassium concentrations at temperatures below those normally necessary for diagenesis will promote illitization (Weaver, 1968). This may account for the presence of illite mixed layering in montmorillonite even in seabed sediments.

Powers (1967) theorized that diagenesis commenced at approximately 6000 feet and continued at an accelerating rate to a depth of 10,000 to 12,000 feet where it abruptly terminated "...where there is no discrete montmorillonite," resulting in a mixture of pure illite and some mixed layer illite/montmorillonite (Figure 2-12a). Powers explains this by use of the Arrhenius equation indicating that the reaction rate would be temperature dependent. This implies an assumption that the reaction has zero order both in reactants and products. In fact, the reaction is at least first order both in the reactants and in the decline in montmorillonite and potassium ions. Initially this is in excess but later declines due to both reaction and flushing by desorbed water, which would require declining rather than accelerating rates of reactions.

Burst (1969) provided a more rigorous evaluation of reaction rates, resulting in a modified dewatering curve (Figure 2-12b) in which diagenesis commences at a greater depth than in Powers' model. His curve shows rapid acceleration to maximum at approximately the same depth, and then declines asymptotically. This model more closely fits the reaction kinetics of the process in which temperature tends to accelerate the process, while depletion of reactants and increasing pressure act to slow it. Burst, however, chooses to explain the curve not on the basis of kinetics but by proposing a two-stage process of rapid loss of the second layer of hydration water followed by a lower rate of loss of the first layer in illite conversion.

While dismissing Burst's two-stage model as being inconsistent with X-ray data, Perry and Hower (1972) also disagreed with Powers' reaction rate curve. They did, however, confirm Powers' reaction model, that montmorillonite-to-illite conversion involves the loss of both hydration water layers simultaneously, but proposed a new two-stage process (Figure 2-12c, 2-12d).

Perry and Hower proposed the rapid onset of diagenesis, reaching its maximum at a depth somewhat shallower than in Burst's proposal. This process, involving the collapse of random montmorillonite layers to illite with subsequent water desorption and expulsion, would decline with depth (in agreement with reaction kinetics) until approximately 65 percent of the montmorillonite layers have been converted to illite. At this point, they proposed a second rapid increase in diagenetic rate resulting from the transition from random to ordered interlayering.

This process appears to peak and decline rapidly, ending with approximately 20 percent montmorillonite, 5 percent water of hydration and 2 percent interstitial water. Below this point, all authors agree that a slow process combining physical and diagenetic dewatering continues at a rate normally too low to be discernible through commonly drilled intervals at this depth.

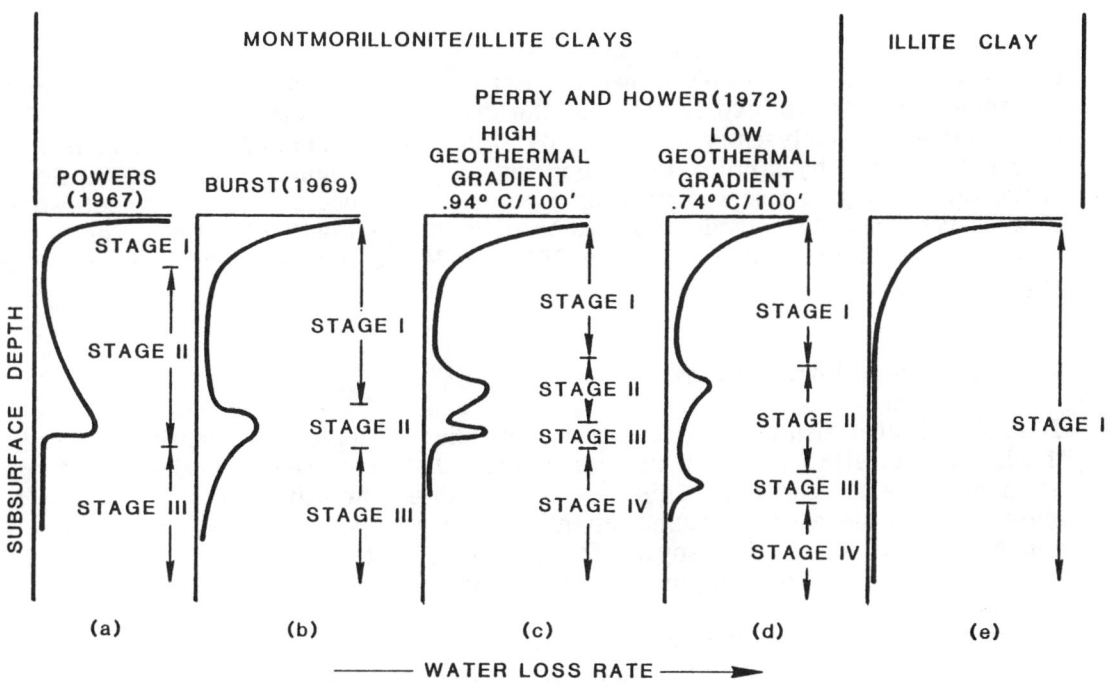

Figure 2-12. Theoretical Dewatering Curves

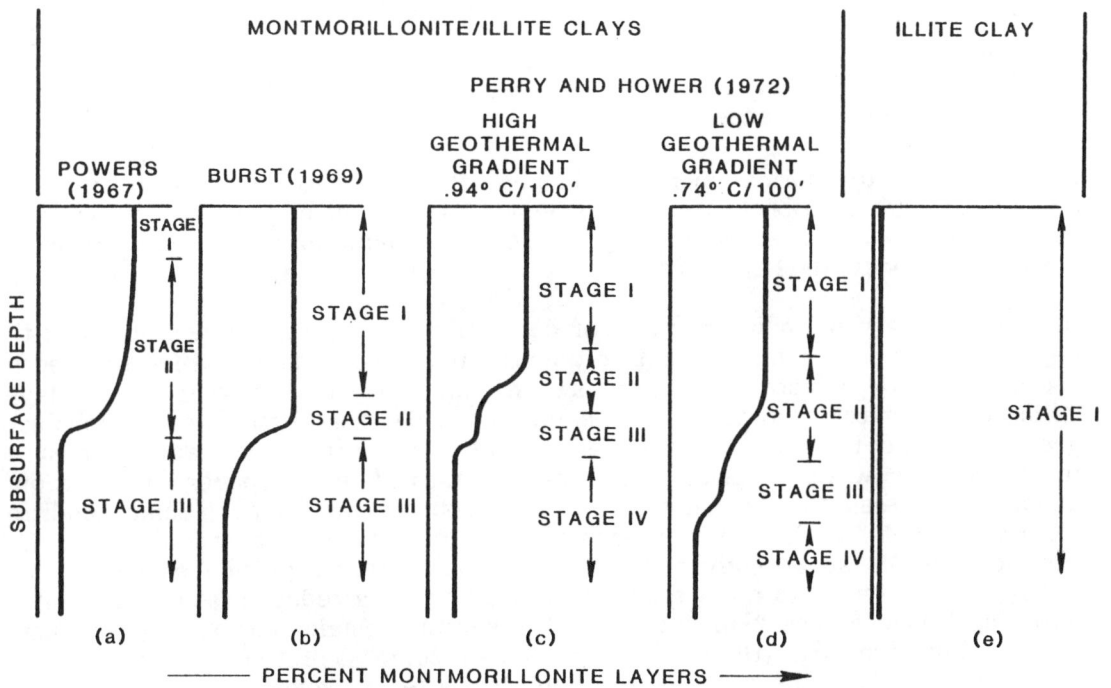

Figure 2-13. Theoretical Montmorillonite Composition Curves (derived from Figure 2-12)

The various dewatering models appear radically different when considered in terms of the water loss rate curves shown in Figure 2-12. However, when the percentage montmorillonite contained is plotted, the difference becomes less marked. Since montmorillonite analyses provided the data upon which each of the models is based, it can be seen why similar data, bearing in mind common minor mineralogical variations, may give rise to differing conclusions (Figure 2-13). All models show a progressive, possibly episodic conversion of montmorillonite-to-illite spread over a temperature range of approximately $80^\circ C$ to $200^\circ C$ with maxima grouped around $100^\circ C$ to $110^\circ C$.

2.30 Abnormal Pore Pressure

The role of interstitial fluid pressure in clay diagenesis is not directly addressed in the literature, although most authors suggest that increased pressure acts to slow down the rate of reaction. This may be explained theoretically. Clay illitization involves a net increase in volume (change of compact water of hydration to the less compact liquid state), thus any factor acting against this increase in volume (e.g. fluid or overburden pressure) may act to impede the rate of reaction.

Practical observations also confirm this hypothesis. Under normal diagenetic conditions (i.e. short of metamorphism), the reaction never goes to completion and the total destruction of montmorillonite. This suggests that the reaction rate declines not simply due to depletion of reactants but due to some additional factor — pressure from depth of burial at which some limiting point, with reactants still remaining, effectively reduces the reaction rate to zero. Where geothermal gradient is lower and diagenesis begins at greater depth, it is seen that the conversion of montmorillonite to illite with depth is at a rate slower than that for which lower temperature alone would account. Again it can be seen that depth of burial, hence pressure, is having a negative effect on reaction rate. Pressure thus can reduce the rate of the conversion reaction, but it cannot alone reverse or even (in theory) entirely halt the process which under conditions of burial is an irreversible chemical change.

In addition to the normal increases in pore fluid pressure with depth of burial, abnormally high pore pressures act to reduce the rate of clay montmorillonite-to-illite conversion. Such pressure may be externally imposed or may result from the conversion process itself.

In a sedimentary basin where rates of deposition and subsidence are uniform, the depth of diagenesis remains relatively constant. At this depth (which is fixed by the geothermal gradient), with pressure and chemistry having some effect, layer collapse and water desorption will begin. Since desorbed liquid water occupies a greater volume than interlayer water of hydration, either permeability must be increased for water expulsion or the interstitial pore spaces will become abnormally pressured. In many cases, it is probable that some median condition results in which partial expulsion occurs, pore fluid pressure abnormality rises, and reaction rate is accordingly reduced to balance water expulsion capacity. The result will be similar to a low geothermal gradient — a reduced rate of illitization with depth (see Figure 2-14b, 2-14c). The section remains abnormally pressured, but equilibration with fluid expulsion causes some depletion. Such abnormal pressures, in which the pressure gradient increases to some abnormal value and

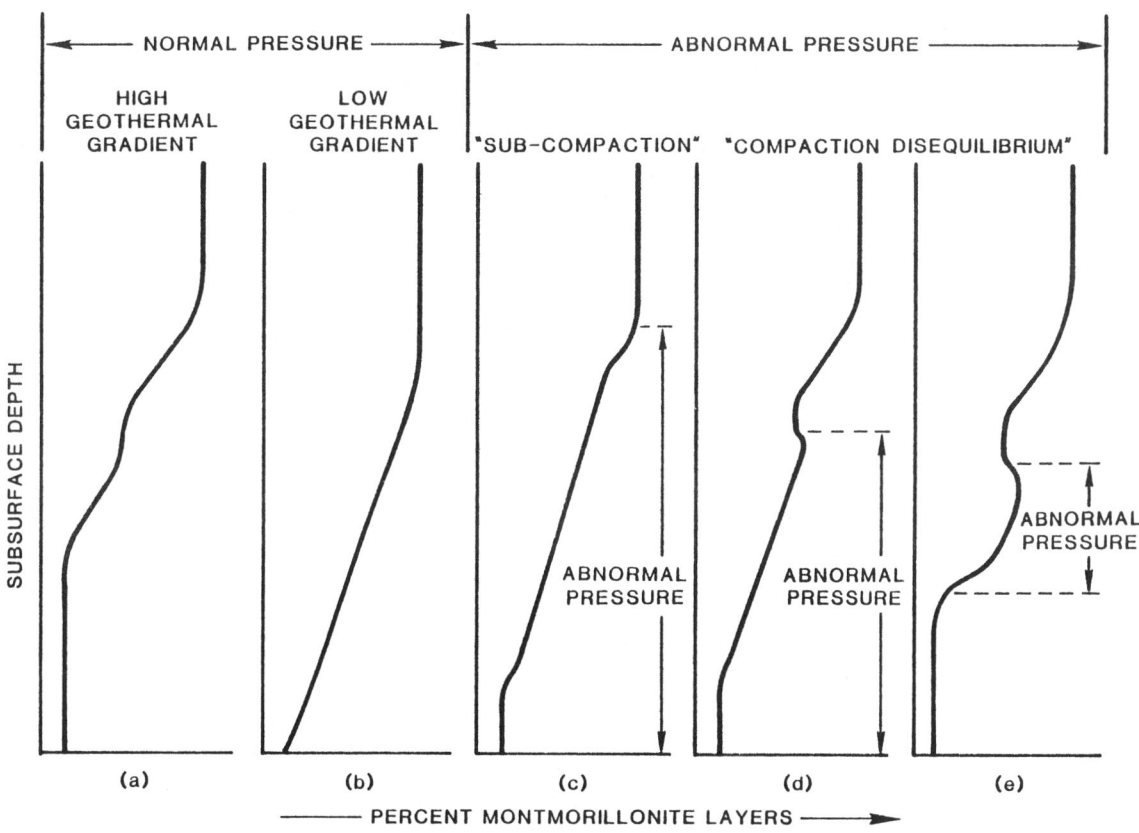

Figure 2-14. Effect of Abnormal Pore Pressures on Montmorillonite Composition

remains relatively constant or slightly declining with depth, have commonly been attributed to "subcompaction." The term "compaction" in this use is a misnomer, and "subdiagenesis" may be more appropriate.

Neither sedimentation rate nor source is totally uniform throughout a section. Certain zones, due to changes in sediment rate or type, may have reduced dewatering ability. Transition from a normally pressured, normally dewatered zone to an abnormally pressured zone may be marked by an apparent diagenetic reversal (Figure 2-14d, 2-14e). This increase in montmorillonite content over a depth interval does not mean that montmorillonite-to-illite conversion has been reversed, but that the conversion is occurring at a retarded rate in the abnormally pressured zone which therefore has a higher montmorillonite content than the younger, normally pressured zone above it. Such abnormal pressures, in which the pressure gradient increases to some high value over a long interval (reflecting a transitional sediment change) and then may revert to normal, are commonly attributed to "compaction disequilibrium." Again, this is a misnomer since compaction is not the process involved and the formation has in fact reached an equilibrium reaction rate dictated by temperature, pressure and chemistry.

In some cases such overpressures are found to contain (often near their upper limit) thin, dense bands of mineralized, often calcitic, material. It has been the practice to refer to this band as a "seal" or "cap rock," implying that its lack of permeability gave rise to the overpressure. For such a band to have been precipitated during sedimentation is environmentally improbable, and no mechanism exists for its precipitation prior to clay diagenesis. However, calcium ions desorbed by the clay are dissolved in desorbed water and flushed from the clay. In normal dewatering the material is carried to aquifers where it serves as a source of cementing minerals. In an overpressured clay, normal dewatering is greatly impeded. Some upward water and ionic migration will occur with favoring of the smaller mobile ions. Accumulation of larger ions may be sufficient to allow precipitation and the formation of a mineralized zone. This process may, therefore, only be justified as a possible side effect and accompaniment to abnormal pressure formation, but never as its cause.

Not all abnormal pore pressures are diagenetic in origin. Pressure abnormalities are encountered in clay sediments which are too young or too cool for diagenesis to be a factor. Such overpressure must result from a failure in the initial physical dewatering of the sediment. If interstitial water content and porosity cannot be reduced with burial, increased sediment loading rapidly produces highly abnormal fluid pressures. Unlike overpressures generated in diagenesis, these are unstable and unrelated to the other physico-chemical parameters of the system.

Physical dewatering of shallow sediments is a simple hydrodynamic process and, under normal circumstances, the fluid conductivity of a clay sediment is capable of completing it. For dewatering to be stopped, some major event in the sedimentary sequence is required. Such an event must radically and over a short time interval reduce the vertical permeability of the sediment.

Such an event could be a sudden increase in the rate of sedimentation, especially if accompanied by an increase in the content of the clay sediment of nonexpandable clays. These clays, containing no interlayer water, would more rapidly dewater and lose permeability than the underlying expandable clays. Even in these circumstances, the presence of laterally extensive permeable interlayers or thin sands would allow dewatering to continue by lateral fluid expulsion.

The overpressured zone, effectively isolated by more dewatered sediments above and below and lacking lateral permeability, even without further compaction, would suffer further increases in fluid pressure with increased burial. The production of methane by biogenic activity and early kerogen diagenesis would further overpressure the pores. In addition, temperature increases with burial would increase the pressure of the fluid contained in the fixed pore volume.

These increased pressures cannot be sustained by the formation, thus hydraulic flowpaths must be formed. This commonly results in increased vertical permeability, allowing fluid expulsion sufficient to maintain constant fluid pressure. Evidence of this is shown by massive clay sections of several thousand feet which, though apparently lacking effective permeability, exhibit a relatively constant pore pressure gradient throughout, indicating that some means of pressure redistribution has has been achieved.

If pore pressure cannot be relieved in this progressive manner, fluid pressure will rapidly increase with burial and temperature increase, approach the level of

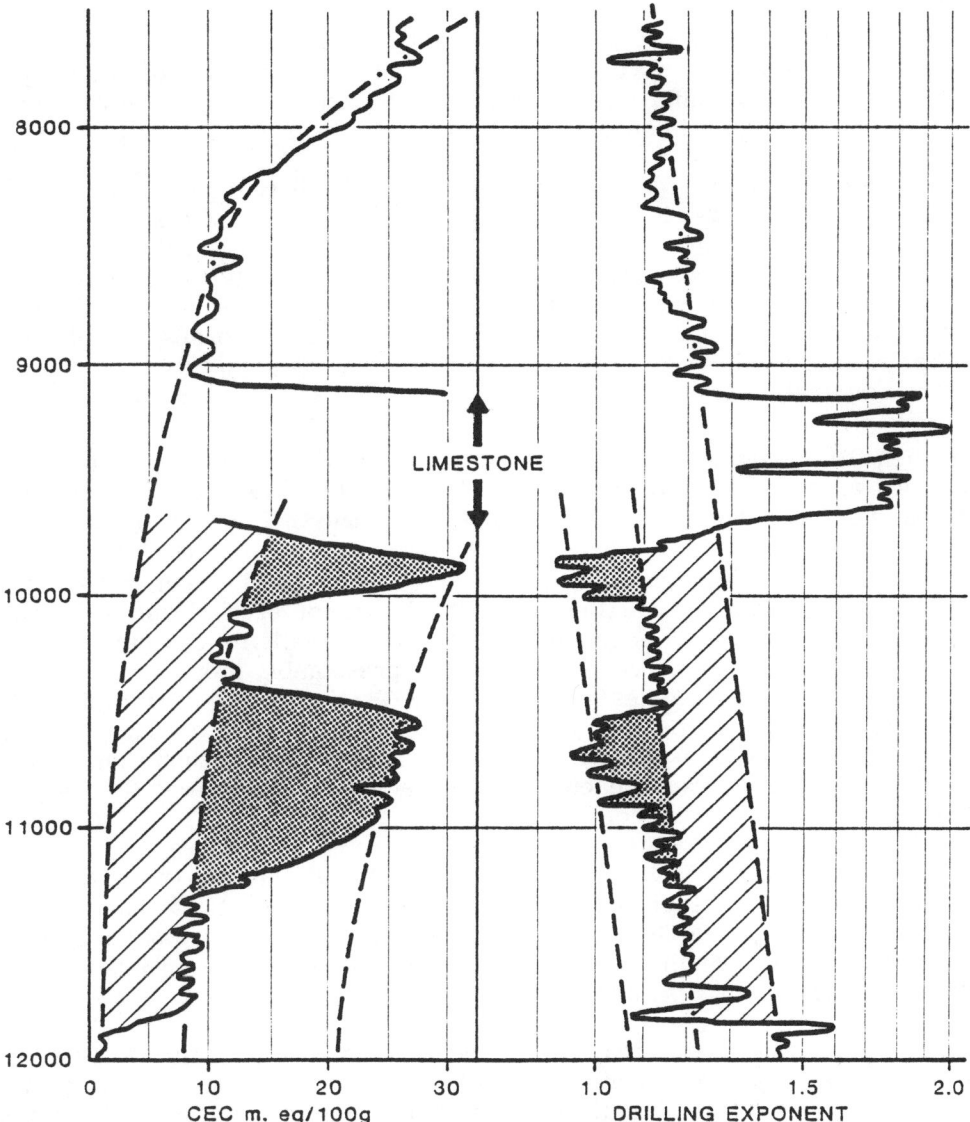

Figure 2-15. Overpressure Indication from C.E.C.

overburden load, and result in major formation deformation or dislocation and resultant fluid distribution.

It is predominately as a tool in pressure evaluation that the use of C.E.C. measurement on shale cuttings has been advocated in the past. Figure 2-15 shows a casebook example of this. Both C.E.C. and a drilling exponent confirm that, below the massive limestone at 9500 feet (obviously, in this case, sufficiently massive to confirm primary sedimentary origin), diagenetic rate has been slowed and abnormal pore pressure resulted. At 9800 and 10,500 feet, zones of extra pressure abnormality occur, possibly due to periods of higher sedimentation rate or higher sediment montmorillonite content.

This example shows the ideal pressure-indicating ability of C.E.C. measurement (compare with Figure 2-14e). However, in cases where overpressure development occurs prior to diagenesis or commences early in diagenesis (Figure 2-14c), indications will not be so definite. Where lack of uniformity in sediment source leads to radical variation in montmorillonite content with depth, interpretation may become difficult or result in "false alarms." This reinforces the rule that, like all other pressure-indicative tools, C.E.C. should be used only as one component of an overall evaluation.

Even when C.E.C. cannot give a definite indication of the presence of overpressure, it can act as a forewarning tool, indicating the likelihood of the occurrence of overpressure. For example, in Well 2/11-1 (Figures 2-7 and 2-8), Karlsson et al (1978) interpret the decrease in montmorillonite content in the Upper Miocene as being indicative of a change in sediment source (decreased vulcanicity). They also conclude a climatic deterioration and higher rates of erosion and sedimentation. Regardless of this sophisticated interpretation, a wellsite evaluation of C.E.C. measurements would have a rise in C.E.C. and presumably montmorillonite content with depth. Kerogen studies of the Lower Tertiary (see paragraph 2.31) also indicate the section to be gas prone. This, coupled with the absence of sand intercalations, would have suggested the strong possibility of subnormal dewatering and abnormal pore pressure deeper in the section. The Lower Tertiary in this section is indeed overpressured.

In older rocks, high C.E.C. values indicate a highly montmorillonitic rock. A progressive decline in such values accompanies the onset of diagenesis, again indicating that a mineralogy and depth in which abnormal pressures are possible have been reached. Such forewarnings should be taken as a signal for special vigilance and care in drilling and formation evaluation activities.

Apart from pressure evaluation, C.E.C. has a major use as a formation evaluation tool. As previously mentioned, it provides a quantitatively determined variable through lithologies which visually may prove difficult or impossible to distinguish. While clay mineralogy or C.E.C. alone may prove insufficient to explain a variation of the data, a characteristic marker or discontinuity may still be identified and used.

The value of C.E.C. measurement in identifying potential source beds is similarly important. Petroleum source beds are more likely to occur in montmorillonite clays, and to have matured sufficiently to produce hydrocarbons, they must have reached temperatures similar to those resulting in illitization. For the hydrocarbons to have been efficiently expelled from the source rock, illitization must have proceeded to completion at a normal temperature-controlled uniform rate. Evidence of all these factors is provided by C.E.C. measurement.

2.31 Source Potential

The organic component of the clay rock bears evidence of source and sedimentary history. This is controlled by the vulnerability of organic material to biogenic and chemical degradation in the sedimentary environment. Thus it may be said that, to survive this environment and be preserved, the relocation of organic detritus from the aerobic environment (necessary for its development) must of necessity be rapid.

The classic example of an environment satisfying these conditions is the euxinic, Black Sea or fjord, stagnant or silled basin. Upon death, the rich faunas of the oxygenated surface water sink to the anaerobic bottom water where biological activity, other than that of sulfate-reducing bacteria, is suppressed. Such environments yield sediments of high organic content (exceeding 10 percent) but are not typical or common in the sedimentary record (Hunt, 1979). Thus to explain the varying degrees of survival of organic material requires consideration of the factors operating in more common sedimentary environments.

Biological activity requires an aerobic environment and organic nutrients. Where either of these is lacking, such as in arid continental or marine abyssal plains, biological productivity is inhibited and sediments are organically poor. Where oxygen is present in excess, such as in high-energy coastal environments, productivity may be adequate but the high oxygen content stimulates both biogenic and chemical oxidation. Conversely, in nutrient-rich organically productive environments such as coastal swamps, biological activity may exceed the ability of free oxygen to oxidize organic matter. Similarly, zones of minimum oxygen content (governed by seawater temperature and hydrostatic pressure) in marine basins, while not conducive to biological growth, are most favorable to preservation of organic detritus derived from elsewhere in the basin.

Assuming a constant supply and degradation rate of organic material, the organic richness of the eventual sediment is in inverse proportion to the rate of mineral sedimentation. Hence environments with high rates of sedimentation (such as deltas) are organically poor. However, where excessively low rates of sedimentation occur, organic material is exposed for long periods on the seabed and in the aerobic top few feet of sediment where degradation rates are high.

In general, organic-rich sediments form in or adjacent to areas of high biological production in quiet water, the oxygen content of which is naturally low or depleted and where sedimentation rate is intermediate and uniform (Dow, 1978). In the terrestrial environment these conditions are typified by a subsiding coastal swamp, forming an almost pure organic sediment of vegetal material an in acidic, biologically inhibited water. In the marine environment the best conditions are those suited to the formation of fine-grained argillaceous and carbonate rocks. Carbonates, in which both mineral and organic sediment have a common source, tend to form in an open marine environment — resulting in a uniform composition and distribution of marine material. Clay rocks have variations in quantity and quality of organic material which may be of marine or terrestrial origin. This variability is enhanced by the clay mineral's ability to precipitate polar organic molecules dispersed or dissolved in water. Expandable clay minerals, montmorillonites, are capable of accommodating charged organic molecules within their interlayer spaces, thus concentrating and protecting them from further degradation. Organic concentration in calm water is therefore reduced and transferred to the precipitating clay minerals.

After burial and removal from the aerobic environment, the organic component of the clay rock consists of original biopolymers, some simple monomers resulting from biochemical and chemical degradation, and geopolymer complexes resulting from interreaction. The compounds are present in solution and suspension in interstitial water as amorphous or even organically structured aggregates or, most importantly, bound ionically at active interlayer sites within expanded montmorillonite clay minerals.

The diagenetic processes resulting in the conversion of organic material to hydrocarbons begin almost immediately after burial. It is convenient (Hunt, 1979) to separate and name the processes undergone according to the type of chemical reactions involved. Diagenesis involves catalytic reduction, stripping, and reforming of organic material. Temperature has little effect on the processes, and diagenesis occurs below $50°C$. Between $50°C$ and $200°C$ the major processes occurring are thermocatalytic. Although physical catalysis is required to promote the processes, rates of reaction are strongly temperature dependent. These processes, referred to as "catagenesis," involve cracking, decarboxylating and hydrogenating the products of diagenesis to produce true petroleum hydrocarbons. Beyond $200°C$ the processes become entirely thermal, resulting in the cracking and dehydrogenation to progressively lighter hydrocarbons with the eventual products of methane and graphite. These processes, which are destructive to hydrocarbons, are termed metamorphism. The processes involved in organic catagenesis and metamorphism are chemically and kinetically complex, the rate determining the reaction stage often being a slow one. For this reason, temperature exposure time is often as critical as temperature itself. Thus rocks of more recent burial (not necessarily rocks of less age) may not undergo catagenesis until reaching higher temperatures than those in cooler but longer buried rocks.

It should be noted that the progression from surface microbial to diagenetic, catagenetic, and metamorphic states involves gradual transitions and that the terms are used in a generalized way — less specific than the usage elsewhere in geology.

The reaction products and intermediaries of diagenesis which are no longer organic but not yet petroleum are given the general name "kerogen." The bulk composition of the kerogen, which will strongly affect the nature of the eventual petroleum product, is itself dependent upon the source and environment of organic deposition.

Kerogens may be divided into two major types: (1) sapropelic, originating from lipid organic material, predominately of algal origin and deposited in subaqueous, aneorobic environments, and (2) humic, of vegetal origin, characteristic of aerobic environments (but of course with some oxygenation-limiting factor preventing total decomposition). In petroleum geochemistry it is more common and useful to use three categories based upon the common environmental assemblages of the two source types. The change in bulk composition of these types during burial is shown well in a Van Krevelen diagram as in Figure 2-16.

The hydrogen/carbon ratio in mature hydrocarbons range from 0.45/1 for oil to 0.3/1 for gas. Thus kerogen with high H/C ratios will mature to produce high yields of oil and gas; lower H/C ratio kerogens will be more likely to produce gas and at lower yields.

Type I kerogen, consisting of algal and amorphous material (commonly lacustrine in origin), has an H/C ratio at diagenesis of 1.7/1. Considerable dehydrogenation may occur in diagenesis and catagenesis with accompanying oxygen loss, and still ensure a high enough H/C ratio to give oil and/or gas yields. Type I kerogen provides the richest petroleum sources, but is rare. Type II kerogen, of intermediate origin (usually marginal marine), is an admixture of marine algal and amorphous material with both marine and terrestrial vegetal material. Its initial H/C ratio of 1.4/1 leads to good yields of both oil and gas and provides the most common economic source beds. Type III kerogen, consisting of woody, vegetal material and inert recycled organic material (terrestrial in origin), commencing with an H/C ratio of 1.0/1 or less, yields gas usually in marginal or noncommercial quantities. Preferential tropical weathering of this type of organic material may sometimes favor the preservation of vegetable waxes resulting in low gravity waxy oils. Type IV organic material, which is not a true kerogen, consists of inert carbonaceous material consisting of plant residues and recycled organic debris. It does not yield hydrocarbons.

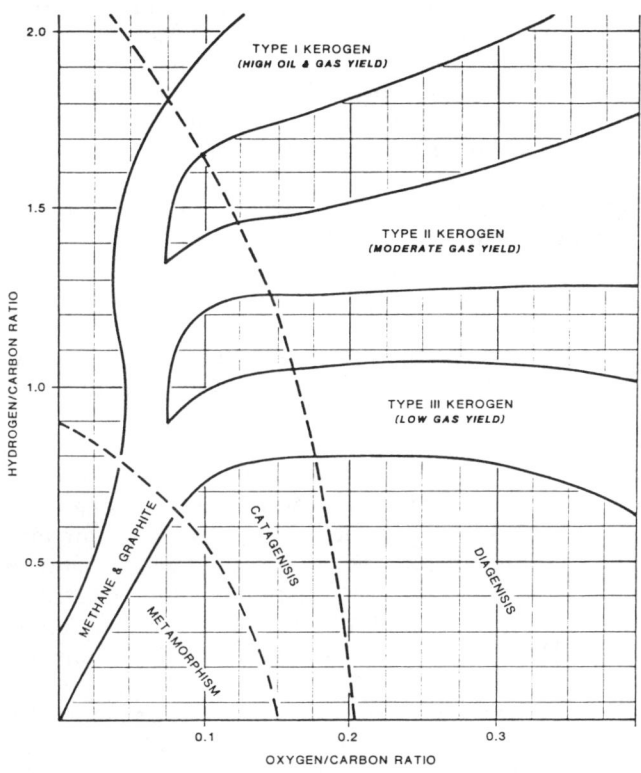

Figure 2-16. Van Krevelen Diagram Showing the Thermal Progression of Kerogens

Although in shallow burial microbial activity is of major importance in the catalytic diagenesis of organic material to kerogen, such activity cannot account for the total reaction, especially at increasing depth where microbial activity declines with temperature. Johns and Shimoyama (1972) confirmed the important role played by expandable montmorillonite clays in the process. They proved both the viability of the catalytic processes and sequential relationship between them and the dewatering and diagenesis of montmorillonite clays.

It is well known that, during settling and sedimentation, surface-active clays filter organic material efficiently from water. This is a simple ionic bonding to interlayer sites of polar organic molecules. Bonding is both cationic at ion exchange sites and of weaker hydrogen bonding type similar to the mobile interlayer water. Montmorillonite clays will therefore remove larger quantities of organic material from the marine environment more rapidly than nonexpandable illite and kaolinite clays.

At shallow depth, during initial physical dewatering, contained organic material would be expected to be expelled along with interstitial mobile interlayer water, leaving the sediment organically impoverished. This does occur in illite clays. In montmorillonite clays, fatty acids and other polar organic material are retained by adsorption at interlayer and surface sites, reducing this depletion effect. Decarboxylation occurs at this same depth and temperature range, resulting in the production of long chain hydrocarbon-like compounds. Johns and Shimoyama showed that the activation energy of this process is lowered by about 18 percent by the catalytic effect of montmorillonite and water.

The depth and temperature range typically associated with clay diagenesis (80°C to 200°C) and the second period of major water expulsion (or second and third, according to Perry and Hower) correspond closely with the range for organic catagenesis (50°C to 200°C). It appears that this correspondence is more than coincidental. The cracking, decarboxylation and hydrogenation processes occurring in the kerogen in catagenesis require acid catalysis. Although their reaction rates are temperature dependent, the reactions and products could not be produced from the effect of temperature alone.

The water of hydration in the final two water layers of montmorillonite has been demonstrated to have acidic character (proton-donating behavior) many magnitudes in excess of that for normal water (Johns and Shimoyama, 1972). This characteristic, due to interaction between the water and exchange cations, provides an active acid catalysis for the carbonium ion reactions long theorized to be the cause of decarboxylation and the cracking of long chain hydrocarbons to shorter chain length, without or with little odd/even chain length preference.

The acid catalysis is effective only while the water is present as water of hydration on the montmorillonite lattice. Once conversion to illite is complete and the water desorbed, reaction is halted. Thus a balance of conditions (temperature, pressure, comparative reaction rate, pore water and kerogen chemistry) must be satisfied to allow the hydrocarbon product to be available in sufficient quantity at the peak of water desorption rate to allow flushing of economically significant quantities of hydrocarbons from the clay rock. Even given this balance, a large proportion of the produced hydrocarbons will remain in the source rock.

Hydrocarbons remaining in the rock or thermally generated later in diagenesis will be subject to thermal expansion and to some volume increase due to molecular rearrangement. These changes will lead to further pressure increases and expulsion by reopening of hydraulic pathways created by water and primary hydrocarbon expulsion. This second stage process is supported by the evidence of microfractures showing two stages of secondary mineralization separated by hydrocarbon residues.

In a nonexpandable clay rock, neither the catalytic mechanism to generate hydrocarbons nor the water efflux to promote their migration is available. Thus the following criteria are favorable to the occurrence of an active petroleum source bed, i.e. one in which hydrocarbons are or have been generated and migrated to adjacent reservoir rocks:

- deposition in moderate depth marine environment
- high montmorillonite mineral content
- sufficient burial to cause diagenesis to illite

The identification and categorization of clay rocks which are active or potential petroleum source beds are as important in an exploration well as identification of potential reservoirs. Studies of kerogens in sediments have indicated good correlation between physical and chemical properties of these and their petroleum potential and maturity. In the past, however, the determination of such properties involved time-consuming and costly laboratory techniques unsuitable to application at the wellsite. The advent of the "Rock-Eval" pyroanalyzer has provided a wellsite tool capable of rapid, reliable source bed characterization. The advantages of the system include more representative and knowledgable sampling and savings in transport and laboratory costs; but most importantly it is available at the time and point of drilling the source bed and is a new and valuable geochemical exploration tool.

The technique of kerogen pyrolysis has been used and described by various authors, including Giraud (1970) and Barker (1974). However, it was Espitalie et al (1977) who first developed a device and method for the programmed pyrolysis of kerogen and the analysis of evolved hydrocarbons and carbon dioxide. The device proved to be amenable to wellsite operation aboard the D/S Glomar Challenger during Legs 48 and 50 of the IPOD program.

Clementz et al (1979) described further developments in the use of the Rock-Eval in a joint project by Chevron Oilfield Research Company, Chevron Overseas Petroleum Company and Exploration Logging Inc. This project resulted in the implementation of various improvements in equipment and technique, resulting in a more reliable and reproducible operation of pyroanalysis in a logging unit in a commercial offshore drilling environment and in the integration of geochemical data with conventional data from the computer-assisted GEMDAS® service logging unit.

The equipment consists of a pyrolysis oven in which a quantity of ground cuttings is heated in a helium atmosphere through a programmed temperature ramp to $550°C$. Evolved gases are analyzed for hydrocarbons and carbon dioxide. In addition, the temperature corresponding to the maximum evolution of hydrocarbons is recorded. As heating progresses (Figure 2-17), two main phases of hydrocarbon evolution are

GEMDAS® is an Exlog registered service mark, standing for Geological and Engineering Monitoring and Data Analysis Service.

seen corresponding to the free hydrocarbons present in the rock, which are volatilized below 300°C (Peak area S1), and the hydrocarbons produced by the cracking of the organic kerogen above this temperature (Peak area S2). Inert (Type IV) organic material will not pyrolyze to procuce hydrocarbons. The method therefore only indicates productive (as opposed to total) organic material. Peak area S3 corresponds to the carbon dioxide produced by the pyrolysis which, in this temperature range, is assumed to derive solely from the organic kerogen. Although carbonates will thermally decompose to give carbon dioxide, in the temperature range used in the Rock-Eval all common carbonates are thermally stable.

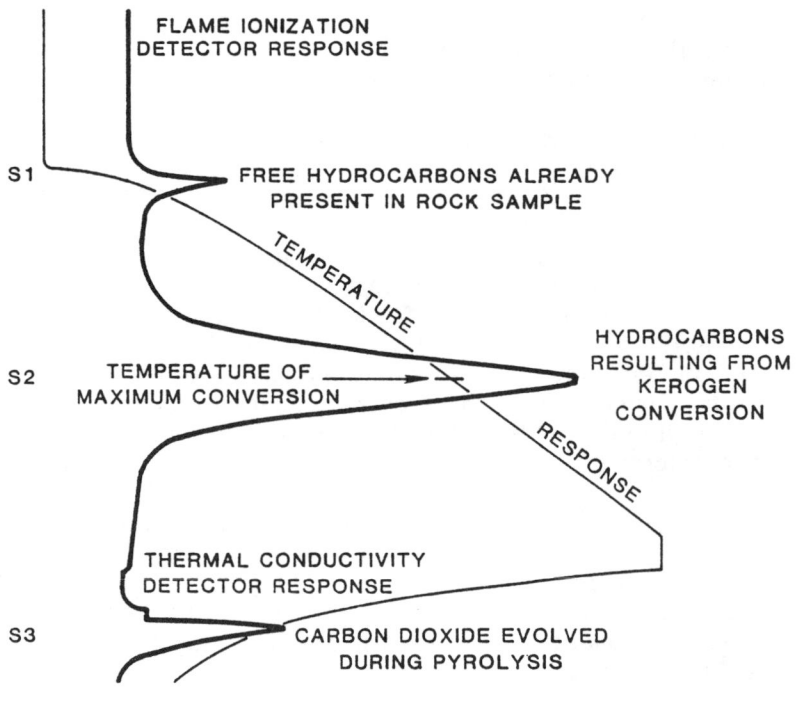

Figure 2-17. Source Rock Evaluation by Pyrolysis

The theory of interpretation is simply that, in order to be a potential hydrocarbon source, the rock must contain large quantities of kerogens. If S2 is greater than 5 mg/g of sample, the rock has good source rock potential. Heavy bitumens may pyrolyze late and contribute to the S2 (kerogen) peak instead of the S1 (hydrocarbon) peak, indicating a falsely low maturity. This effect may be compensated for by comparing S2 peaks from samples before and after solvent washing.

For the source rock to be economically significant, it must be sufficiently mature for mobile hydrocarbons to have been produced and migrated to a reservoir rock. Thus the ratio of free-to-total hydrocarbons, S1/(S1+S2), is indicative of the likelihood of discovering producible hydrocarbons in the vicinity of the source bed, with a higher ratio indicating a higher degree of maturity and mobility of the hydrocarbons.

Finally, it is known that the likelihood of obtaining oil or gas from a source is a function of the relative amounts of oxygen and hydrogen in the organic source material (Figure 2-16) and of the temperature at which it degrades to form hydrocarbons. Source material comparatively richer in oxygen or residues degraded at higher temperatures are more likely to yield gaseous hydrocarbons. Thus the temperature of maximum hydrocarbon generation (Tmax) and the ratio of generated carbon dioxide-to-hydrocarbons (S2/S3) are significant indicators of the type of production to be expected:

$S_2/S_3 < 2.5$: gas-prone kergoen

$S_2/S_3 > 5.0$: oil-prone kerogen

$T(max) < 435$ C: immature, biogenic gas only

435 C $> T(max) < 450$ C: mature oil generation

450 C $> T(max) < 470$ C: mature gas generation

$T(max) > 470$ C: post-mature, thermal gas only

These are, however, generalizations, and a more reliable evaluation of a section's hydrocarbon potential is made by means of a geochemical data log plotting the continuous results of pyroanalysis with depth. In Figure 2-18, the vertical scale has been drastically reduced to allow the representation of a complete section from the immature young sediments via the oil and gas generating zones through to the deeper nonproductive sediments.

It can be seen that with depth the S1 mobile hydrocarbon peak will in general increase at the expense of the S2 kerogen peak as diagenesis, catagenesis and metamorphism of the organic material proceeds. Similarly, the S3 carbon dioxide peak will decrease with depth as decarboxylation and hydrogenation continue. As hydrocarbon maturation and production continue, progressively higher temperatures are required to produce hydrocarbons from the remaining kerogen. Thus the T(max) temperature of maximum hydrocarbon evolution also increases with depth. This temperature, which is a function of the design and programming of the pyrolysis instrument, has no quantitative significance. It does, however, qualitatively indicate the readiness of the kerogen to produce hydrocarbons. This will be greatest in the Types I and II, oil-prone kerogens, and least in the Type III, gas-prone kerogens, and in material which has reached maturity and already liberated hydrocarbons. From these general rules, the empirical dividing line shown on the log may be derived.

In addition, certain rules of interpretation may be drawn, and these are illustrated in the examples shown in Figure 2-18. The immature source bed, shown at A, has

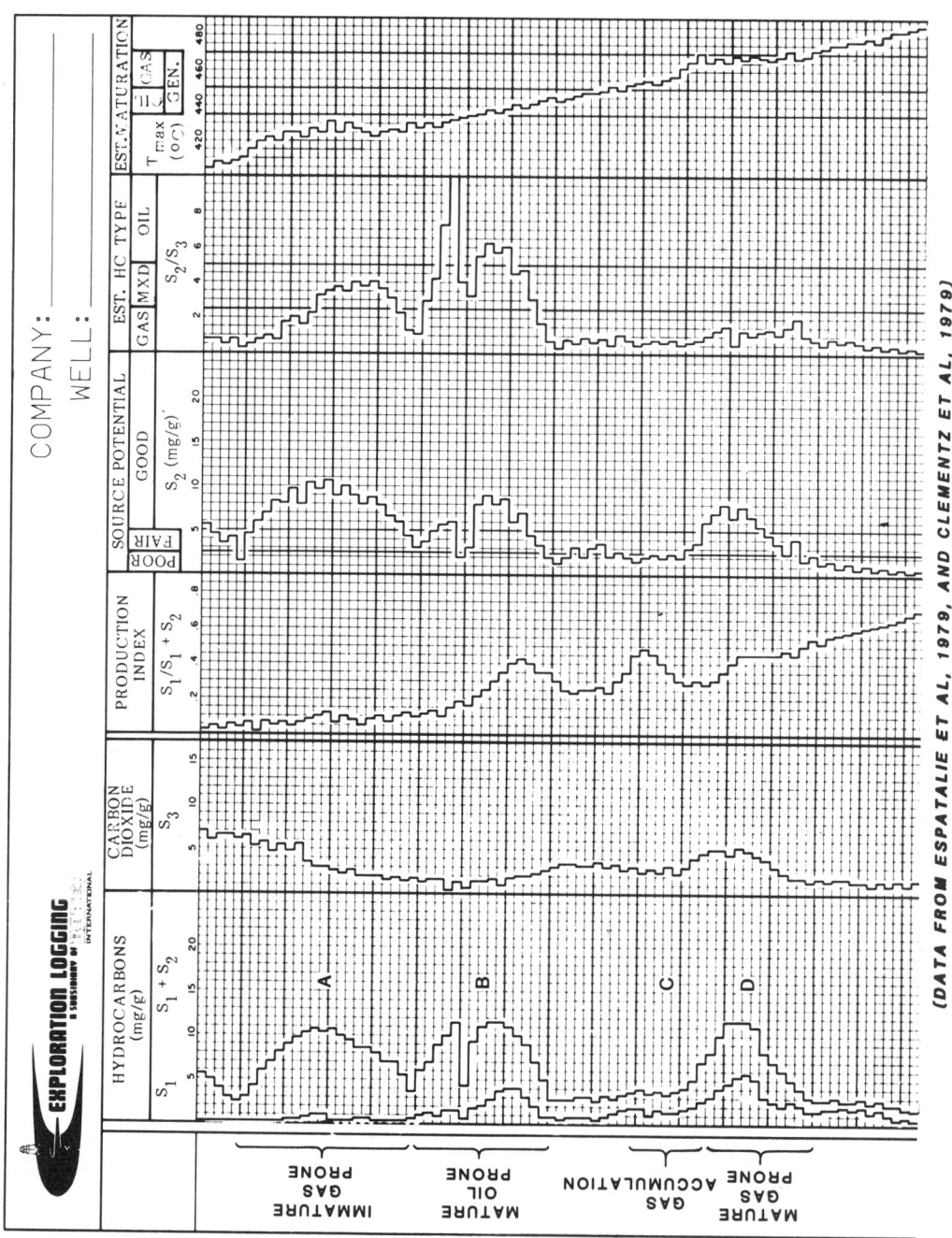

Figure 2-18. Example Pyroanalysis Log Showing Common Events

good source potential shown by S2. The low S1 and production index indicate that hydrocarbon generation has not begun in the zone and this is confirmed by the T(max) which, though increased (indicating a gas-prone source), is too low for generation. Further confirmation would be given by a C.E.C. plot showing that clay diagenesis has not yet commenced. Such immature source beds, while economically unimportant, may be important if they are more deeply buried elsewhere in the basin. At the least they often prove to be useful stratigraphic markers. In addition, high organic content coupled with high montmorillonite and low permeability may act as a warning of the advent of a shallow overpressure, the overpressure being accentuated by the production of biogenic or early diagenetic methane.

At B, an oil source in which petroleum generation has begun is marked by increases in both S1 and S2. The increase in production index lower in this interval indicates that mobile hydrocarbons are accumulating at this point. In contrast, the gas source at D shows some decline in the production index, indicating the depletion of mobile hydrocarbons from the zone. These appear to be accumulating at C which shows little source potential but an increase in mobile hydrocarbons.

Obviously, lithological information is required for the complete evaluation of source or reservoir potential of the hydrocarbon data shown by pyroanalysis. The C.E.C. data, as discussed above, may prove crucial in drawing conclusions about the migration potential of hydrocarbons detected by pyroanalysis.

Clementz et al (1979) reported an unforeseen benefit of continuous wellsite pyroanalysis: stratigraphic boundaries and markers, not lithologically apparent, could commonly be correlated by means of changes in kerogen content and maturity. Like the changes in clay mineralogy discussed by Weaver (1960), the changes are not always immediately understandable; they nevertheless may be recognized and applied.

2.32 Porosity

The porosity in rudaceous and arenaceous rocks is primarily interparticle, consisting of the void space remaining between particles in packing. Once an efficient packing geometry has been established during sediment compaction, further porosity changes will result from introduction or removal of cementing material or solution of grains.

The porosity of perfectly sorted spherical particles (were such a sediment to exist) would be a function of the packing (see Field Geologist's Training Guide (EXLOG, 1985). This varies, becoming more efficient (thus reducing porosity) with increased depth of burial and sediment loading. Theoretical porosities for various packing geometries are shown below:

Geometry	Porosity	Occurrence
cubic	47.6%	young sediments
hexagonal	39.5%	compacted sediments
rhombohedral	25.9%	most sedimentary rocks

Thus the theoretical maximum porosity for a detrital rock is about 26%. In fact, this is much reduced by other factors in actual sediments, and a guide for estimating true porosities is as below:

>15%: Good
10-15%: Fair
5-10%: Poor
<5%: Trace

The logging geologist should not attempt to estimate the porosity percentage. Instead, he should use the terms good, fair, poor and trace or combinations of these.

Although theoretically independent of grain size, it is in fact observed that fine-grained sediments tend to have higher porosity than coarse ones. This phenomenon is a result of the distance and means of transport of the different grain-sized sediments. As has been said before, both physical and chemical weathering and mass sorting of a sediment will proceed as it is carried further from the parent rock. Thus near-to-parent sediments will tend to be coarse, angular and poorly sorted, while distant sediments will be finer, better rounded and sorted and hence approximate to the ideal sediment theorized above. The following table indicates the empirical justification of this supposition by giving the grain size and porosity analyses of various typical clean detrital sediment aggregates.

Rock Type	Composition (% Grain Size, mm)			Porosity (%)
	> 0.5	0.5-0.062	< 0.062	
Silt	2.5	1.9	95.6	46
Silty sand	0.3	65.8	33.9	42
Sand	52.2	36.8	11.0	33
Sand	78.7	20.8	0.5	27
Sandy gravel	88.6	10.7	0.7	25
Sandy gravel	96.3	3.5	0.2	37

It is interesting to note, however, that fine-grained sediments, although having higher porosities in general, may contain less oil. The reason is that most oil sands are preferentially water-wet. The surface of each grain is coated with water, and this water substantially reduces the volume of porosity available to be filled with oil. It is easily demonstrated that the total surface area of the sand grains increases as the grain size decreases. Thus a reduction by half of the grain size of a sediment, though theoretically not affecting porosity, will double the surface area of sediment and hence the irreducible water content.

Sorting of a sediment has a major effect on porosity. Figure 2-19 illustrates sediments in cubic packing. Where there is a mixture of grain sizes, the smaller grains will reside rhombohedrally in the cubic pore spaces of the larger — drastically reducing porosity.

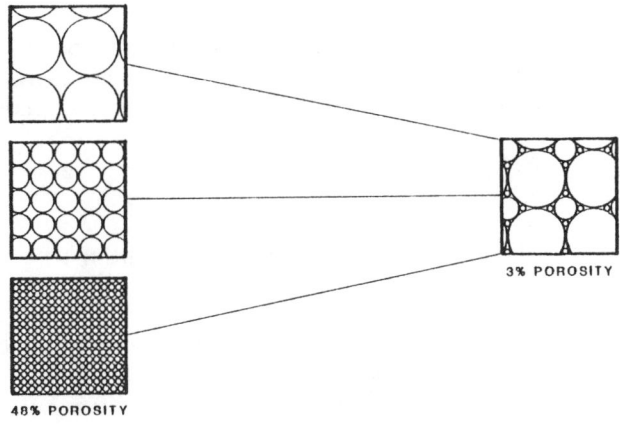

Figure 2-19. Effect of Sorting on Porosity

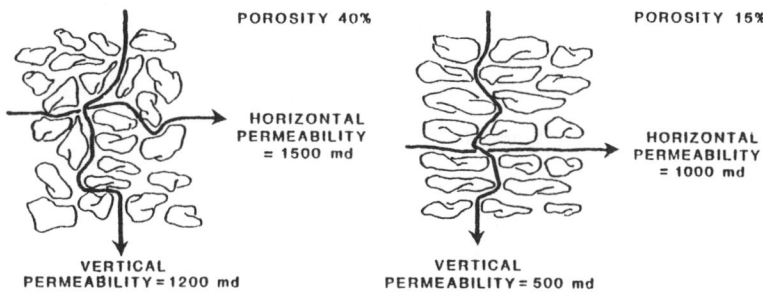

Figure 2-20. Effect of Particle Alignment on Porosity and Permeability

Particle shape also affects porosity. If randomly distributed, decreasing roundness and sphericity of the particles would interrupt the packing geometry in increased porosity. This assumption does not take into account that both sedimentation and compaction tend to force lineation of elongate particles, which may have the reverse effect of reducing the porosity (Figure 2-20).

Cement is the major influence over porosity of rudites and arenites after sedimentation and compaction. The crystallization or solution of cement will result in changes both in the absolute porosity and more importantly the effective porosity (the volume of interconnected void space). The precipitation of cement or secondary mineralization on grains at the pore throats may isolate pores, preventing recovery of fluid.

Since the factors controlling the porosity of granular rocks are comparatively uniform within the rock mass, porosity estimates from cores or even cuttings are reliable and can be extrapolated to the reservoir with some degree of confidence.

2.33 Permeability

The velocity of flow of a fluid through a permeable medium will be a function of

- the viscosity of the fluid
- the temperature of the fluid (since viscosity is temperature dependent)
- the pressure differential across the flowpath
- the cross-sectional area available for flow
- the length of the flowpath
- some constant which may be defined as permeability

Permeability is defined as the constant K in the equation

$$q = \frac{KA \Delta p}{\mu L}$$

where:

- q = flowrate (cc/sec)
- A = cross-sectional area (cm^2)
- Δp = pressure differential (atmospheres)
- L = length of flowpath (cm)
- μ = fluid viscosity (centipoise)
- K = permeability (darcies)

assuming

1. Steady state flow conditions exist.
2. The pore space of the solid medium is 100% saturated with the flowing fluid. Under this restriction the <u>absolute permeability</u> will be considered.
3. That the viscosity of the flowing fluid remain constant. This is untrue since viscosity is temperature- and pressure-dependent, but will be acceptable if the viscosity at the average pressure is taken and if assumptions 4 through 6 hold.
4. Isothermal conditions prevail.
5. Flow is horizontal and linear.
6. Flow is laminar.
7. The fluid is incompressible.

Hence a rock will have a permeability of 1 darcy (1d) when 1 cc of a fluid of 1-cp (centipose) viscosity flows through a 1-cm cube (i.e. 1 cm x 1 sq cm) of rock in 1 second under a pressure gradient of 1 atmosphere (N.B. viscosity of water at $68^{\circ}F$ = 1 cp).

One darcy (d) is a very large permeability in terms of reservoir rocks, and measurements are usually expressed in millidarcies (md or k = K/1000).

The following range of permeabilities is a general guide to permeabilities observed in sedimentary rocks.

Millidarcies	Evaluation
1	tight
1 - 10	fair
10 - 100	good
100 - 1000	very good
1000	excellent

It can be seen from the above argument that many of the assumptions made do not hold true in the reservoir or even in a core sample when a permeability measurement is being made. For this reason a "measured" permeability is never more than a numerical estimate and cannot be directly applied to the behavior of actual fluids in place in the reservoir. When considering the ability of a formation to flow fluids, some consideration must be given to the concepts of relative permeability concerning the type and composition of flowing fluids. See Mud Logging: Principles and Interpretations (EXLOG, 1985) for a discussion of this subject.

In general, the permeability of a granular rock is closely related to its porosity. That is, rocks with good porosity will in general have good permeability. Permeability will also be modified by packing, sorting, particle shape, size and cementation, but not always in the same manner in which porosity is affected.

Permeability is a gauge of the quality of the flowpath through the rock. In general, it depends upon the number of available passages through the rock, the cross-sectional area of those passages and their linearity (fluid flows more readily through a straight pipe than through one with many bends).

More efficient packing geometries will reduce the cross-section area of any flowpath through the rock by bringing the particles into closer proximity. They will also decrease the linearity of the flowpath. Both of these factors tend to reduce permeability. A similar result is seen with decreased sorting which also reduces the area and increases the length of the flowpath.

Particle size affects flowpath area since the size of pore throats (the gaps between particles) is proportional to the radii of the particles. Thus, unlike porosity, permeability tends to increase with increasing grain size.

Particle shape affects permeability in different ways, dependent upon orientation (Figure 2-20). Angular particles in random orientation will, by increasing porosity, increase flowpath cross-sectional area and hence permeability. Conversely, linear orientation of elongate particles reduces porosity and hence permeability. However, elongation also tends to improve flowpath linearity parallel to the long axes of the particles so that permeability in this direction will be improved <u>relative</u> to that in the direction perpendicular to the lineation.

Cementation and secondary mineralization have the effect of reducing permeability by reducing the diameter of pore throats or even by closing them off and reducing the number of available flowpaths.

Argillaceous rocks may have as high or even higher porosities than granular rocks. Interconnection of pores is rare and of poor quality, resulting in both the effective porosity and permeability being very low. For this reason they do not provide good reservoirs other than in the rare circumstance where fracturing provides both permeability and secondary porosity. In this case the rock's reservoir properties are not inherent but are similar to those of a rudaceous clastic, e.g., a boulder scree. It may be thought of as a clastic rock produced in situ at depth rather than at a surface exposure.

Although argillaceous rocks have a low permeability (of the order of 1/10,000,000 md), this is appreciable when considered in terms of geologic time. Thus pelitic source rocks, although low in permeability, may still release maturing hydrocarbons into permeable reservoirs from where they may be produced. It has been suggested that even nonsource rock argillites through which hydrocarbons migrate have a role in the generation of the eventual hydrocarbon composition. The theory is that the mobile precursors of the petroleum migrate from the source bed in an immature form. In migrating through other argillites there is a separating effect, similar to chromatography. The lighter, simpler hydrocarbons migrate rapidly while the heavier, more complex components are retarded. These heavier components may be retained in the argillites or may be "cracked" to simpler fractions under the action of heat using the complex ionic plate structure of the clay lattice as reforming catalysts.

The low permeability of argillites allows them to constitute cap rocks enclosing a petroleum reservoir. However, some diffusion through the low permeability will occur. This explains the old driller's axiom that the occurrence of gas shows toward the base of a shale cap rock is not a discouraging sign but may be an indication that a reservoir below contains hydrocarbons and not water.

It is difficult sometimes to give a reliable estimate of permeability from cuttings alone. However, from the factors which affect permeability it can be seen that, from a complete rock description, including all of the characteristics contained in this section, a reliable conclusion can be drawn. Comparing ditch gas and blender gas also aids in estimating permeability (see <u>Mud Logging: Principles and Interpretation</u>, (EXLOG, 1985).

When estimating permeability in a core, it is better to make general observations of the whole core from the time it is recovered and throughout processing. This will give a more visual estimate than microscopic examination of single core chips. A good general guide (though not necessarily a universal one) for estimating permeability in sandstone cores is as follows:

- Excellent: The core will be poorly consolidated and may fall apart in recovery.

- Very Good: Fluid will be bubbling from the whole core, giving the impression it is effervescing. Bubbling may be so violent as to cause the surface of the core to break up and grains to detach.

- Good: It will be impossible to wipe the core dry. Fluid wiped from the surface of the core with a dry rag will be replaced from within.

- Fair: Core can be wiped dry, but after a period will become wet again.

- Tight: Drilling fluid on the surface of the core will dry in the air without wiping.

Porosity and permeability are the two most important characteristics of a potential reservoir rock. The logging geologist can judge the efficacy of his or her final written description by reviewing it from the point of view of a geologist who has not seen the sample, and deciding whether that person would draw the same conclusions about porosity and permeability from the written observations as they would from seeing the sample. This is the acid test for a sample description. As was said earlier, interpretive comments may be added to a sample description but it is more important to list all direct observations and allow the recipient to make his own interpretations and deductions.

2.34 Hydrocarbon Shows

A description of any visible liquid or solid hydrocarbons is essential in all lithology descriptions, be they source, reservoir or simply remnant hydrocarbons indicating that migration has passed this point in the section. A complete review of hydrocarbon test procedures is contained in Mud Logging: Principles and Interpretation, (EXLOG, 1985).

3
CARBONATE ROCKS

3.1   **CLASSIFICATION**

For the purposes of this manual, the term "carbonate rocks" will be used for those marine rocks consisting predominately of calcium carbonate (calcite or aragonite) and calcium magnesium carbonate (dolomite). Other carbonate minerals do occur in sedimentary rocks, but they are rarely rock-forming minerals. These are included in Figure 2-10, Accessories Occurring in Sedimentary Rocks, and discussed in Section 4, Chemical Rocks.

It is difficult to classify carbonate rocks due to the complexity of the sources and types of their occurrence. A simple classification may be made on the basis of origin, as follows:

- Clastic
    - Lithoclasts: fragments of preexistent rock or sediments
    - Bioclasts: fossil fragments or faunal accumulations

- Chemical
    - Lutaceous: lime mud or micrite
    - Crystalline: sparry limestone or dolomite

Unfortunately, the solubility of carbonates and their tendency to recrystalize (resulting in the loss or distortion of sedimentary textures) make this system ambiguous. To describe a rock in terms of its current physical state may be misleading, but to classify it by origin only may lead to false conclusions as to its current characteristics, most notably its potential as a reservoir.

A rock classification is a self-consistent system of rock names and modifiers allowing the most significant rock characteristics to be summed into the minimum number of words, commonly a single name with modifying prefixes. Each classification differs in the characteristics upon which it is based. Similarly, in different regions or applications, differing rock characteristics may be judged to be most significant.

Once a classification has been adopted there are also benefits of consistency and mutual intelligibility in retaining it. Service companies usually adopt and work with whichever classification is favored by the clients for whom they work.

A rock name or classification is no more than a label attached to a discrete group of rock characteristics. If complete lithological evaluation and description are performed and reported, the selection of a rock name or classification from any system is a logical and simple process.

Various classification systems are currently used by the petroleum industry. Each has virtues and weaknesses and commonly each has special value within specific applications. The most common classifications are reviewed in paragraphs 3.2 through 3.22. The Exlog geologist should provide a complete geological description following the procedures explained in paragraph 3.23 through 3.69. Following this, a rock name may be selected from the client-specified classification.

3.2 PETTIJOHN

This system recognizes only two basic types of carbonate rock:

- Granular
- Crystalline

The first generally applies to limestones only, but may be used for dolomites where the results of textural alteration are insignificant. Further subdivision is based on the energy of the environment of deposition. Textural differences are attributed to high- or low-energy deposition (Figure 3-1). The term "crystalline limestone" (or dolomite) is applied to low-energy deposition. The term may be modified by crystal size and amount of contact between crystals.

3.3 DUNHAM

This classification consists of a modification and extension of that of Pettijohn's. The basis of classification remains that there is fundamental textural distinction between sediment laid down in low- and high-energy environments. Under high-energy conditions, little or no fine material remains at the site of deposition due to the effect of current removal. Low-energy, calm-water conditions are typified by more dominant amounts of mud. The system discriminates between grains (>0.02mm) and mud (<0.02mm). In terms of the microscope power commonly available at the wellsite, this may be considered as visible particles and indistinguishable particles. No discrimination is made in the system between lime and clay muds (Figure 3-1).

3.4 ARCHIE

Archie's classification does not concern itself with lithological or genetic character of the rock and, in reality, is not a rock classification system at all. It is a porosity classification system which is of value in the oil industry but seldom used elsewhere. Even within the industry it is normally used as a modifier to some other more rigorous classification, commonly Dunham (Figure 3-1).

Archie's system defines two types of porosity:

- Matrix-defined (or "invisible")
- Nonmatrix-defined ("visible" at x10 power or lower)

Subdivisions within these types are allocated combinations of letters and numbers as described in paragraphs 3.5 and 3.6.

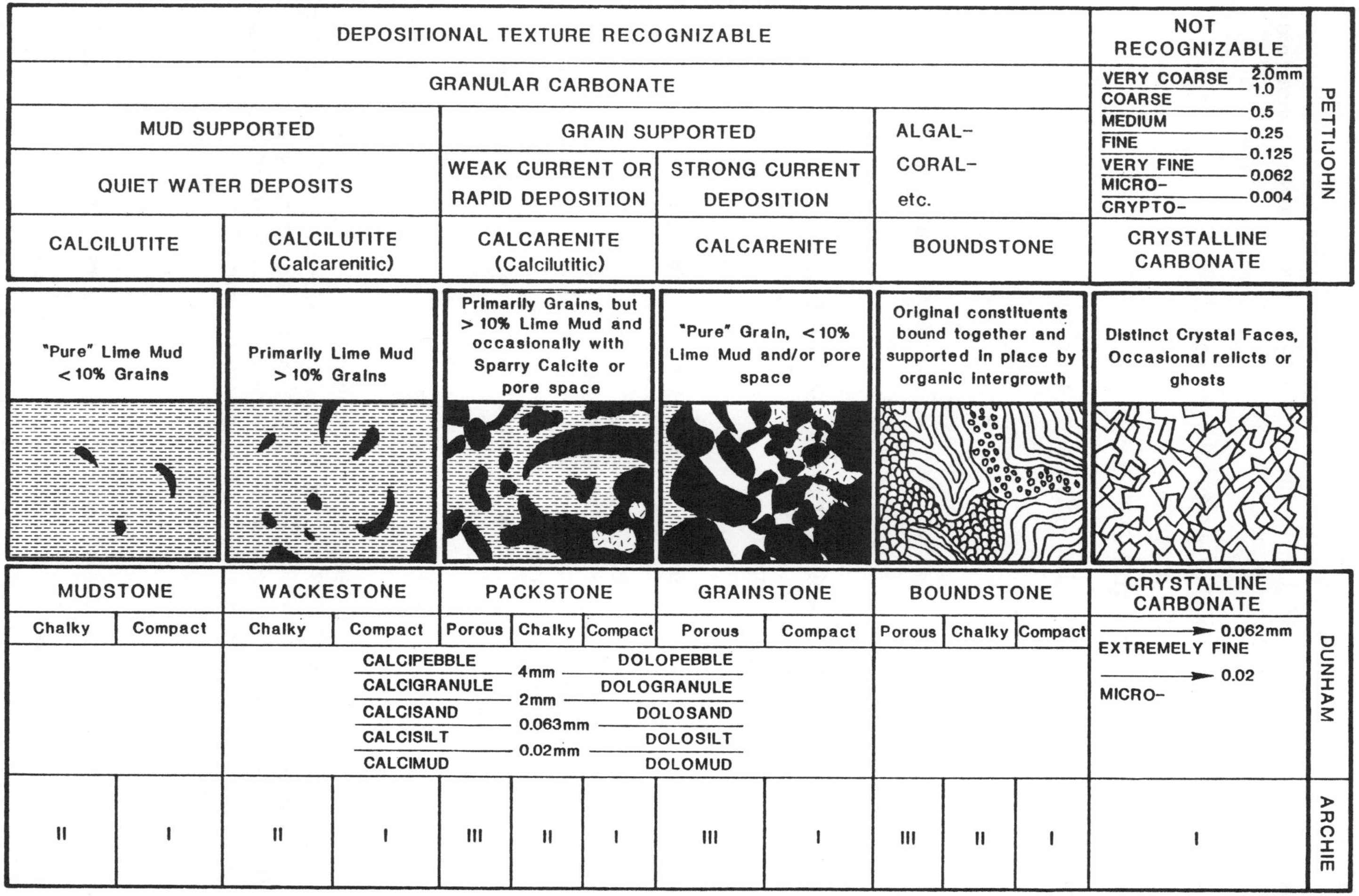

Figure 3-1. Pettijohn/Dunham/Archie Carbonate Classification

3.5 Matrix

This is porosity which is not visible to the naked eye or x10 power magnification but may be inferred from inspection of the matrix.

   I: Compact hard rock with sharp or conchoidal break
   II: Chalky or earthy frangible rock
   III: Sucrosic crystalline rock with little induration

A combination of two or more of these classes is acceptable.

3.6 Visible

This is visible pores or vugs greater than 10 microns (visible with x10 power magnification):

   A: No visible pores
   B: Pores visible with x10 power (0.01mm < pore < 0.1 mm)
   C: Pores visible with the naked eye (0.1mm < pore < 2mm)
   D: Vugs or cavities (pore > 2mm)

Combinations are acceptable, such that a hard indurated crystalline rock may be described as a I.A, and a weakly cemented chalky mudstone may be described as a I/II.B/C.

A combined Dunham-Archie classification with genetic modifiers can prove to be an acceptable wellsite classification tool, as it provides most of the information required by the drilling and petroleum engineers. It does not lend itself easily to geological investigation, although geologists who have worked with it extensively have much to say in its favor.

3.7 FOLK

Folk's classification attempts to segregate major classes of carbonate rocks into groups analogous to the detrital rocks (or terrigenous rock in Folk's terminology). For the three major constituents of detrital rocks, clastic or resistate particles, clay or silt matrix, and cement or secondary minerals, Folk invoked the analogous limestone constituents allochems, micrite and sparry calcite (Figure 3-2).

3.8 Allochems

Allochemical constituents, or allochems, include all particulate material in a limestone. Although formed within the depositional basin (rather than transported to it from outside) by chemical or biochemical precipitation, they have undergone transport within the basin to the site of deposition.

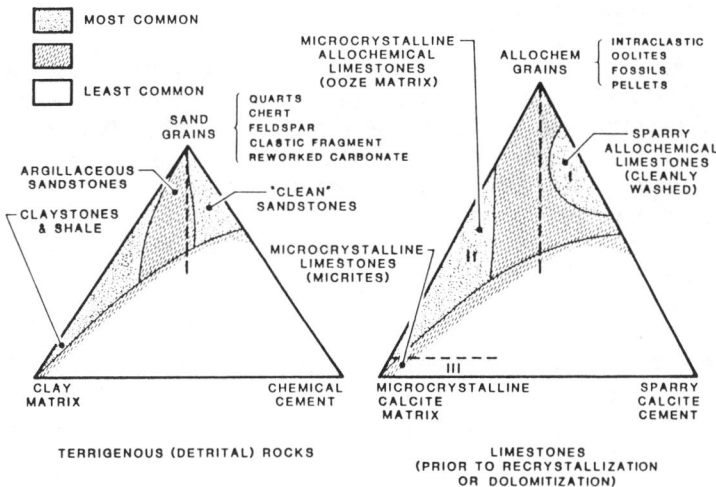

Figure 3-2. Analogous Occurrence of Limestones and Detrital Rocks

3.9 Intraclasts: These are fragments of previously deposited and partially consolidated carbonate sediments which have been subjected to reworking and redeposition. The fragments are abraded to rounded or irregular shapes with the abraded margins which may cut across bedding, fossils, oolites, pellets or even earlier generations of intraclasts. Intraclasts result from submarine erosion caused by storm waves, slides, mild upwarp or tidal and wave action on exposed mud-cracked flats. Their sizes range from very fine sand to boulder, and they are well rounded and vary from spherical to discoidal.

Folk specifically excluded fragments of consolidated limestone eroded from outcrops on a land area adjacent to the basin. These he named <u>calclithites</u> and classified as detrital psammites.

3.10 Oolites: In this classification, oolites were considered to be any particle showing radial and/or concentric structures. Pisolites, grains of crinkly laminae formed by accreted layers of blue-green algae, and superficial oolites having a large nucleus and only a thin oolite coating are included.

3.11 Fossils: This includes all sedentary and transported, intact and fragmentary material with the exception of coral or algal structures growing in situ and forming relatively immobile resistant masses.

3.12 Pellets: These are rounded, spherical to elliptical aggregates of microcrystalline calcite ooze devoid of any internal structure and possibly of faecal origin. In any one rock they show uniform shape and size and may be discriminated from intraclasts by this uniformity.

3.13 Microcrystalline Calcite Ooze

Micrite, the more common modern term for this sediment, is the carbonate equivalent of mud. It consists of gray-white, substranslucent microcrystalline to ultrafine-grained calcite (0.001 to 0.004 mm). It is orthochemical, that is, formed by rapid chemical or biochemical precipitation in seawater, settling to the bottom at its point of formation. Later weak current action may result in drifting. This slight drifting leaves no structural trace in the sediment which is therefore considered to be orthochemical (precipitated in place without significant transport). Subsequent strong current action on compacted sediment may result in the formation of allochemical intraclasts.

It has been proposed that a proportion of micrite is "dust" from abrasion of shell debris. As such, this would be allochemical and could not fall within the above definition. There is doubt as to what proportion of micrite is of this origin, and there is no method for distinguishing the two types. However, it is believed that, except in rare circumstances, micrite resulting from shell abrasion is quantitatively insignificant. Since all micrite behaves in a similar manner in sedimentation and burial, the distinction is not made in Folk's classification.

Micrite, in addition to being the prime constituent of lithographic (compact, featureless) limestone, also forms the matrix of poorly washed limestones and the source material of pellets, intraclasts and some oolites.

3.14 Sparry Calcite

Spar forms as a pore-filling cement precipitated in place within the sediment. It tends to be clear and transparent and averages 0.02 to 0.10 mm in crystal size.

3.15 Limestone Categories

3.16 Type I: These limestones (designated as sparry allochemical rocks) consist chiefly of allochemical components cemented by sparry calcite cement. These rocks are environmentally equivalent to clean cemented sandstones and conglomerates (orthoquartzites) in that solid particles (allochems) have been heaped together by currents powerful or persistent enough to winnow away any microcrystalline ooze that might otherwise produce a matrix. Like their detrital equivalents, the formation of (sparry calcite) cement is governed by limits of packing geometry:

- A minimum of allochems must be present as a supporting framework before cement can form.

- Packing geometry of the allochems fixes the amount of pore volume available. Pore space will always be available for spar to grow and will limit the maximum amount of growth.

3.17 Type II: These limestones (designated as microcrystalline allochemical rocks) consist of allochemical components in a micrite matrix, indicating weaker current conditions. Sparry calcite cement is rare since little initial porosity is present in the sediment. These rocks are environmentally equivalent to the range from argillaceous sandstones to sandy claystones.

3.18 Type III: These limestones (designated as microcrystalline rocks) correspond to the detrital claystones, mudstones and shales. They consist almost entirely of micrite and result from deep water or sheltered shallow water deposition. Some microcrystalline rocks show cavities, caused by burrowing organisms or bottom perturbation, filled with sparry calcite. Such rocks are considered "disturbed microcrystalline rocks" or "dismicrite."

3.19 Type IV: These limestones (designated as biolithite) are biohermal structures growing in situ and forming a resistant mass. Such rocks have no equivalent among the detrital rocks. Limestones consisting of broken fragment or detritus are not considered to be Type IV limestones since, by being broken and transported, the fragments are allochems.

3.20 Limitations

The complete classification (shown in Figure 3-3) has the advantage over Dunham in being able to combine lithological and genetic information in a single rock name. A second term must be added where grain size is significant, e.g., allochemical rocks with intergranular porosity. Similarly, accessory allochems and detrital material may be accommodated in terms of prefixes.

NOTE

> In Folk's classification, the term "calcilutite" does not refer to lime mud or micrite, as in Pettijohn's, but to the smallest fraction of allochemical components. As such, it becomes either ambiguous or useless since such particles either do not exist or cannot be detected (see paragraph 3.13).

The major weakness of this classification, as with others, is that no accommodation is made for textural modification by recrystallization and dolomitization. Primary dolomite rocks do occur analogous to the types of limestone discussed above and may be accommodated in the classification by use of the prefix "dolo". Most dolomite rocks, however, are secondary in origin, produced by the replacing of calcite, with a resultant loss, modification or distortion of structure. Recrystallization (for example, the conversion of micrite matrix to spar cement) may not be readily recognizable without thin-section analysis under high magnifications not available at the wellsite. The strictness of Folk's classification may force selection of a rock name unrepresentative of the rocks's current state and unjustifiably environment-specific on the basis of wellsite evidence. Alternatively, the wish to categorize a rock type in a single rock name may lead to an ungainly assemblage of prefixes and suffixes.

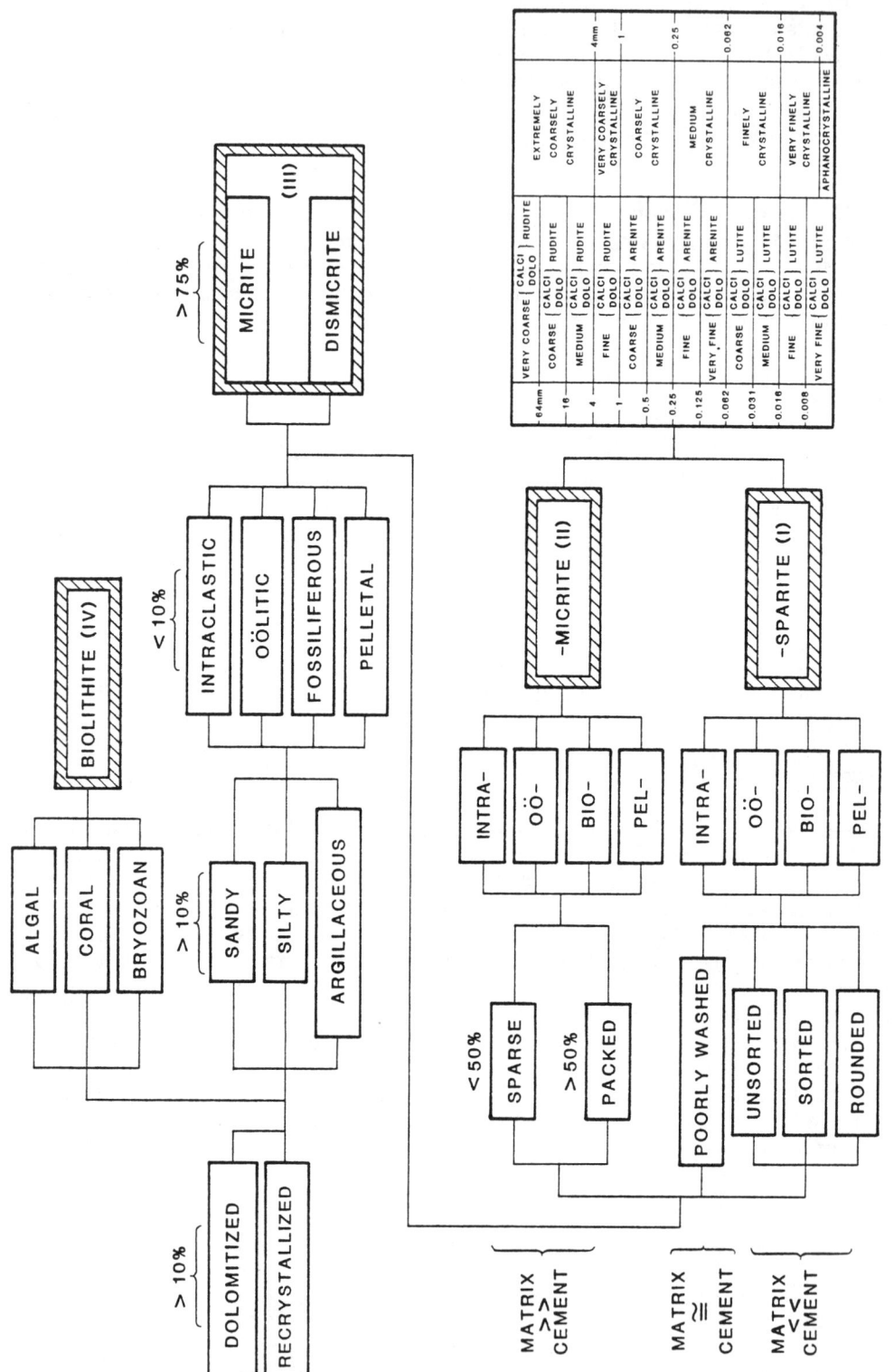

Figure 3-3. Folk Carbonate Classification

3.21 LEIGHTON AND PENDEXTER

This classification system is based upon textural differences and with particular respect to limestones showing evidence of mechanical deposition. These limestones were labeled by the authors as being "clastic textured" (in terms of the classification established in Section 2 in this manual this would be, more correctly, "detrital textured").

The various rock types within this clastic-textured group owe their characteristic appearance to the occurrence and quantity of four textural components. As with Folk's classification, two further components are needed to bring biohermal and crystalline limestones into the system. The six components then are as below:

- Grains
- Micrite
- Cement
- Pores
- Organic frame builders
- Recrystallization

The types of carbonate grain considered in the classification are as follows:

- Detrital grains, and preexistent limestone fragments
- Skeletal grains, and shell fragments
- Pellets, without internal structure
- Coated grains, oolites and pisolites
- Lumps, grain aggregates

The classification (shown in Figure 3-4) is based upon the presence and proportions of these features. Grain size, cementation and crystallization may be added as prefixes to the rock name.

GRAIN/MICRITE RATIO	% GRAIN	GRAIN TYPE					ORGANIC FRAME BUILDERS	NO ORGANIC FRAME BUILDERS
		DETRITAL GRAINS	SKELETAL GRAINS	PELLETS	LUMPS	COATED GRAINS		
9:1	~ 90%	DETRITAL LIMESTONE	SKELETAL LIMESTONE	PELLET LIMESTONE	LUMP LIMESTONE	OÖLITIC, PISOLITIC, ALGAL, etc. LIMESTONE	CORALLINE, ALGAL, etc. LIMESTONE	CALICHE TRAVERTINE TUFA
1:1	~ 50%	DETRITAL MICRITIC LIMESTONE	SKELETAL MICRITIC LIMESTONE	PELLET MICRITIC LIMESTONE	LUMP MICRITIC LIMESTONE	OÖLITIC, PISOLITIC, ALGAL, etc. MICRITIC LIMESTONE	CORALLINE, ALGAL, etc. MICRITIC LIMESTONE	
1:9	~ 10%	MICRITIC DETRITAL LIMESTONE	MICRITIC SKELETAL LIMESTONE	MICRITIC PELLET LIMESTONE	MICRITIC LUMP LIMESTONE	MICRITIC-OÖLITIC, PISOLITIC, ALGAL, etc. LIMESTONE	MICRITIC-CORALLINE, ALGAL, etc. LIMESTONE	
		←——————————— MICRITIC LIMESTONE ———————————→						

Figure 3-4. Leighton and Pendexter Carbonate Classification

Limestone Type According to Energy Index	Limestone Sub-type	Mineralogy	Texture			Characteristic Fossils	
			Size	Sorting	Roundness	Abundance and Complexity	Phylum, Associations and Preservation
QUIET I Deposition in quiet water	I₁	Calcite Clay (15 to 50%) Detrital quartz (<5%)	Microcrystalline carbonate (<0.06mm) or any size fossil fragments in a microcrystalline carbonate matrix (matrix <50%)	Matrix: good Fossils: poor	Original fossil shapes; angular fragments if broken	Barren to moderately fossiliferous, simple assemblages	Crinoids; echinoids; bryozoans (fragile branching types); solitary corals; ostracods; thin-shelled brachiopods; pelecypods, and gastropods. Foraminifera; sponge spicules; tubular, encrusting, and sediment-binding algae; fecal pellets of bottom scavengers. Common fossil associations are crinoid-bryozoa assemblages, bivalve shell assemblages, Foraminifera assemblages (predominantly planktonic). Many fossils are whole and unbroken and are not mechanically abraded. Any fragmentation of fossil material probably is due to disarticulation upon death, to predatory (boring, opening and breaking) activity and scavenger activity or to solution.
	I₂	Calcite (predominant) Clay (<15%) Detrital quartz (<5%)	Any size fossil fragments in microcrystalline matrix (matrix <50%)	Matrix: good Fossils: moderate to good		Moderately to abundantly fossiliferous, simple assemblages (Coquinoid limestone)	
	I₃		Microcrystalline matrix (>50%) Micrograined to medium-grained clastic carbonate and terrigenous material				
INTERMITTENTLY AGITATED II Deposition in alternately quiet and agitated water	II₁	Calcite (predominant) Clay (<25%) Detrital quartz (<50%)	Microcrystalline matrix (>50%) Coarse to very coarse-grained clastic carbonate and terrigenous material	Matrix: good Clastic material: poor to good	Clastic carbonate material subangular to rounded. Roundness of terrigenous clastics is principally a function of size. Oolites may be present	Barren to moderately fossiliferous, moderately simple assemblages	Characteristic fossils and fossil associations are similar to Type I limestones and also may be more or less rounded by wave action. Scattered fragments of fossils from rougher water environments may be present.
	II₂						
	II₃		Interbedded microcrystalline carbonate and any size clastic. Microscale rythmic bedding	Sorting good within individual lamina		Barren to moderately fossiliferous, moderately complex assemblages	

Figure 3-5. Energy Index Classification of Limestones for Sediment and Diagenetic Carbonates

SLIGHTLY AGITATED III Deposition in slightly agitated water	III₁	Calcite (predominant)	Micrograined clastic carbonate (<0.06mm) predominates	Matrix: good	Barren to sparsely fossiliferous, simple assemblages	Echinoderm, bryozoan and bivalve shell debris; Foraminifera; encrusting algae. Common fossil associations are Foraminifera-abraded bivalve shell fragment assemblages. Fossil materials comminuted from larger fossil structures are well abraded by wave and current action.
	III₂	Calcite (predominant) Detrital quartz (up to 50%)	Very fine-grained clastic carbonate (0.06 to 0.125mm) predominates	Clastic material: moderate to good	Barren to moderately fossiliferous, simple assemblages	
	III₃		Fine-grained clastic carbonate (0.25 to 0.5mm) predominates	Matrix: poor Clastic material: moderate to good	Barren to abundantly fossiliferous, simple to moderately complex assemblages	
MODERATELY AGITATED IV Deposition in moderately agitated water	IV₁	Calcite (predominant)	Medium-grained clastic carbonate (0.25 to 0.5mm) predominates	Matrix: poor	Moderately to abundantly fossiliferous, simple to moderately complex assemblages	Crinoids, echinoids, bryozoans, brachiopod and pelecypod shell fragments, colonial coral fragments, stromatoporoid fragments (Silurian and Devonian predominantly) tubular algal fragments, colonial algal fragments (rare), encrusting algae. Common fossil associations are similar to associations of Types I, II and III, or they are mixtures of these associations. Fossil materials are generally broken and abraded.
	IV₂	Detrital quartz (up to 50%)	Coarse-grained clastic carbonate (0.5 to 1.0mm) predominates	Clastic material: moderate to good		
	IV₃		Very coarse-grained clastic carbonate (1.1 to 2.0mm) predominates		Moderately to abundantly fossiliferous, moderately complex to complex assemblages	
STRONGLY AGITATED V Deposition and growth in strongly agitated water	V₁	Calcite (predominant)	Gravel-size clastic carbonate (rock fragments and fossil material 2.0mm) predominates	Matrix: poor Clastic material: poor to moderate	Sparsely to moderately fossiliferous, complex assemblages	Crinoids; echinoids; encrusting bryozoans; thick-shelled brachiopods, pelecypods, and gastropods; colonial coral fragments; stromatoporoid fragments (Silurian and Devonian predominantly; colonial algal fragments, rudistid fragments (Cretaceous predominantly). Fossil associations are similar to Type IV. Fossil materials are generally broken and abraded.
	V₂	Clay (<5%) Detrital quartz (<25%)	Gravel-size conglomeratic or brecciated carbonate (2.0mm). Tectonic breccias excluded	Matrix: poor Clastic material: poor	Clastic material subrounded to well rounded. Pisolites may be present	Barren to sparsely fossiliferous, complex assemblages	
	V₃	Calcite	Not applicable	Not applicable	Not applicable	Abundantly fossiliferous, simple assemblages (fossil colonial growth in place)	Colonial corals, stromatoporoids, colonial algae (principally the Rhodophyta or red algae and some genera of the Cyanophyta or blue-green algae).

Figure 3-5. Energy Index Classification of Limestones for Sediment and Diagenetic Carbonates

3.22 PLUMLEY

The "Carbonate Energy Index Classification" developed by Plumley, Risley, Graves and Kaley is shown in Figure 3-5 (pages 3-10 and 3-11). Again, the classification is based upon the faunal and textural qualities of the rock. The system is based upon primary (mainly biogenic) rock features which, it is claimed, are readily distinguishable under wellsite conditions and when recrystallization and dolomitization are present. As the name implies, the system is based upon a quantification of energy level and environments of the basin and therefore inevitably links a rock name with a purported depositional environment. While some geologists find this acceptable, most prefer a purely observational classification system.

3.23 DESCRIPTION

The purpose of a rock classification is to allow the summing up of a rock's major constituents and properties in the minimum number of words. It also serves to allow the arrangement of rock types into significant groups or series. The awarding of a rock-type name is an interpretive function; that is, having reviewed and considered all observations, the geologist chooses a suitable name. That name commonly brings with it implications beyond the observed facts (for example, the environment of deposition or mode of transport). This is especially true for the limestone classifications reviewed above and especially risky for rocks as complex and as imperfectly understood as carbonates.

In the following paragraphs a procedure is outlined for the classification of carbonates in an entirely descriptive way. Terms and names used are based upon uninterpreted observation and in general are derived from Pettijohn; hence, they are compatible with most of the above classifications.

The first task is to describe the rock. Having done this, characterization can be made according to whichever of the above classification systems, or some other, is favored by the client oil company. A rock description on a log should contain sufficient observed fact to allow the reader to make his own decision. The addition of an interpretive title may be helpful but the reader must be able to see what evidence supports it. If the client does not specify a classification method, the Exlog recommended descriptive method can stand alone without further addition.

It is a common fault of new geologists to consider the name as being the most important part of a rock description. Although always written first in a rock description, it is in fact the sum of all rock characteristics and should be the last item recorded by the describing geologist. When all other characteristics of the rock have been seen and described, a correct rock name, within the appropriate classification system, will be apparent.

A good example of the danger of using a rock name without a rock description is the term "Marl" for which a brief review of the literature yields the following definitions:

- Calcareous clay
- Clay with shell fragments
- Impure limestone
- Calcareous silt
- Clays containing glauconite
- Dolomitic clay
- Unconsolidated sandy clay with shell fragments
- Calcareous sand
- Fresh water limestones

"Marl" is an extreme case, and one that should never be used by a logging geologist, unless the term is part of an oil company's classification system and a definition has been provided but other, more subtle examples exist. Confusion of rock names can be avoided only by giving a complete description.

3.24 SAMPLE PREPARATION

One of the difficulties of evaluating carbonates at the wellsite is that many of the significant features are either too large or too small for ready identification using cuttings and the regular binocular microscope. Indeed, porosity is often so small as to require electron micrography or so large and randomly distributed as to defy estimation even from cores.

Special techniques, not normally required for detrital rocks, are sometimes used when investigating a carbonate reservoir. Special training will be provided for logging geologists using such techniques. A brief section is included here (paragraphs 3.25 and 3.26) as an introduction to thin sections, peels and other methods.

3.25 Cores

When studying core chips, or even whole core, etching of the rock surface is an aid in identifying texture. For core chips this can be achieved by immersing them in dilute (10%) hydrochloric acid for 10 seconds (dolomite and impure carbonates may require up to 5 minutes and gentle heating). After immersion, the chip should be rinsed and dried. To examine a small section of a large piece of core, place a drop of acid onto a horizontally held section of the core. When reaction ceases, add a second drop of acid and repeat until five drops have been used.

If a diamond saw is available at the wellsite the core can be slabbed, horizontally and vertically, and fresh surfaces exposed. Slabbed surface should be polished using a lapping wheel. If a lapping wheel is not available, polishing can be accomplished by grinding on a glass plate with coarse carborundum grinding compound and finishing, on a separate plate, with fine "1000 grade" grinding compound. After polishing, the surface should be etched by immersion in very weak acid (1%; one part 10% HCl to nine parts distilled water) for 10 to 15 seconds.

Use of a petrographic microscope to examine horizontal and vertical thin sections from cores yields evidence regarding porosity, texture, fauna and recrystallization. A small piece (a few millimetres thick) of slabbed core is taken and the flattest side polished using lap-wheels or progressively finer grinding compounds on glass plates. The polished side is mounted onto a microscope slide using "cooked" Canada balsam. After cooling, the reverse side of the specimen is ground using fine rubbing compound to a thickness of 0.03 mm (although thickness is not so critical as when a polarizing stage is being used). A glass cover slip is cemented in place to protect the thin section.

An alternative to thin sections, though not quite so useful, is the preparation of acetate peels. These can be produced easily and in large sections up to six inches or more. Two methods are available: (1) using a solution of cellulose acetate, or (2) the more practical method using a cellulose acetate film. The liquid form is made up as follows:

Parlodion	112 g
Butyl Acetate	1000 cc
Amyl Alcohol	40 cc
Xylene	40 cc
Ethyl Ether	12 cc
Castor Oil	12 cc

As an alternative to this rather offensive mixture, acetate films are available from 0.002 to 0.020 inches in thickness. These are softened with acetone and applied directly. Softening time varies with film thickness, and experimentation may be necessary.

Slab, polish and etch the core as described above, and then dry and leave it to cool. If the liquid acetate is used, pour it onto the rock surface and allow it to set for 24 hours. Alternatively, pour acetone onto the etched surface and roll a piece of acetate film, matte side (if one exists) down across the acetone-wet area. Drive the excess acetone away along the advancing film-rock contact, maintaining a slight "bead" of liquid along the contact by tilting the rock slightly toward the direction of film advance. The film should not be pressed down, and at least 1 inch of margin of film should be left untouched by acetone around the softened area. Another method is to bow the film into a "U" shape, touch the center of the surface with the base of the "U," and gradually flatten from the center outward.

After setting (15 minutes for the film), gently peel the acetate from the rock surface, soak it in dilute hydrochloric acid for 10 minutes or as necessary to remove adhering rock fragments, then rinse and dry. The resulting peel is mounted between glass sheets and may be used as a photographic negative and enlarged up to x50 (or even x80 in special cases) for the grain size and textural studies.

3.26 Cuttings

Limestone cuttings must be etched, rinsed and dried before attempting any estimate of texture, grain size and porosity. Only color can be reliably seen in a wet sample. Etching can be achieved by immersion as with core chips.

While etching cuttings, test for acid reactivity. The reaction of a carbonate rock with 10% hydrochloric acid is the most common and most commonly misused of the tests in wellsite geology. To have any real value it must be applied systematically. Dumping acid into a full tray of sample gives only the grossest of estimates and may often be misleading.

There are three recommended methods of testing with acid; they are given here for comparison. Use whichever you decide, or all three, as applicable in any particular lithology.

CAUTION

Never apply acid to a sample under the microscope and never place cuttings which are effervescing under the microscope. Droplets of acid carried by effervescence may cloud and eventually damage the objective lens. Some microscopes have a protective transparent cover over the objective lens. If this is in place you may observe effervescence. Always check!

Test 1: The simplest and least time-consuming test is simply to observe the cuttings samples while etching. Keep a dish of dilute (10%) hydrochloric acid available in the sample processing area (covered when not in use and replaced when discolored or at least once per tour), and beside it a dish of clean water. Drop cuttings into the acid dish, retrieve with tweezers, and transfer them to the water dish for rinsing. The reaction in acid indicates the calcium carbonate content of the sample and also the etching time required (violently reactive cuttings should be retrieved immediately, slowly reactive or unreactive cuttings may be left as long as 5 or 10 minutes). Reaction rates may be quantified as follows:

- Limestone (90 to 100% $CaCO_3$): Sample reacts violently, floats on top of the acid and moves about on the surface.

- Dolomitic limestone (50 to 90% $CaCO_3$): Sample reacts immediately but only moderately and moves about within the acid between the bottom and the surface.

- Calcitic dolomite (50 to 90% $CaMg(CO_3)_2$): Sample reacts slowly at first but accelerates to a continuous reaction after a minute or two with some bobbing at the bottom of the dish.

- Dolomite (90 to 100% $CaMg(CO_3)_2$): Very slow reaction; bubbles evolve one at a time with several seconds passing between the release of each bubble.

Argillaceous, anhydritic, siliceous, strongly indurated limestones, or cuttings with oil staining preventing wetting with acid all may react more slowly than indicated above. Where this is suspected, a better test is the test tube method where, if necessary, crushed cuttings may be used.

Test 2: Place one or two (crushed or whole) cuttings into a micro-test tube or cut bottle and cover with an excess of dilute hydrochloric acid. Allow the reaction to proceed to completion (or as long as time allows), warming if necessary. Please note that gentle warming only is to be used. Strong heating may cause the acid to boil and bump, and even dilute acid may give a serious burn when boiling. Observe the reaction and categorize as follows:

- Limestone: Reaction is instant and violent; the specimen floats on the acid and dissolves completely within minutes, leaving the acid frothy.

- Dolomitic limestone: Reaction is moderate but begins instantly and is continuous. With slight warming, the reaction becomes violent.

- Calcitic dolomite: Reaction is weak at first but accelerates after a few minutes. With warming, reaction may become moderately strong.

- Dolomite: Reaction is slow and hesitant and will eventually (up to one-half hour) halt entirely. Warming may speed reaction a little at first, but not appreciably. Acid becomes milky.

Test 3: The best systematic test is with use of a cut dish. After etching and rinsing, inspect the cuttings under the microscope. Select and place several cuttings individually in the depressions of a cut dish. If variation in texture, color or other properties is seen in the sample, select cuttings representative of each type. Then add dilute acid to each depression, and note the initial reaction. Allow the reaction to proceed to completion, with warming or replenishment of acid as needed. Reaction rates are as described in the test tube method above. After completion of the reaction (the logging geologist need not wait for this but may carry on with his other duties while checking the reaction periodically), pour off any surplus acid and carefully rinse the residue with distilled water from a wash bottle; allow to air dry. Inspect the residue under the microscope for extra textural and compositional evidence. For example, oil, clay or silica residues are visible; dolomite textures are emphasized; anhydrite may be tested for with barium chloride (yielding an insoluble residue of white barium sulfate), and the consolidation of the residue (whether it retains the original cutting's form or has collapsed) gives evidence of composition and structure.

Thin sections of cuttings may also be made by mounting in Canada balsam prior to grinding. These may aid in assessing crystallinity and organic structures.

3.27 ROCK CLASS

The term "rock type" has been so diversely used in the various carbonate classifications, and so strongly related to genetic factors, that here we coin the separate term "rock class" to designate the differentiations of two major classes of carbonate which often can be made conclusively, without supposition or implication, by a brief examination of the sample. The two classes are described in paragraphs 3.29.

3.28 Sedimentary Carbonates

These are carbonates in which the original sedimentary character of the rock is unchanged from that present at the time that lithification began. It should be noted that this does not exclude all rocks in which crystal forms occur. The formation of crystalline cement within pore spaces and the recrystallization of aragonite fossils to calcite, <u>without loss or distortion of particle structure boundaries and spatial arrangement</u>, are accepted processes of lithification.

Within the above definition, rocks composed of mainly reef or shoal-building organisms in their position of growth, if unaltered, qualify as sedimentary carbonates although their actual mechanism of development is not entirely sedimentary in nature. Rock of an evaporitic origin may be sedimentary if precipitation of crystals in seawater has been followed by settling with characteristic sedimentary and bedding structure. If crystal growth has occurred within a bottom or intratidal zone, the rock is essentially diagenetic in type.

3.29 Diagenetic Carbonates

These are limestones and dolomites in which diagenesis has <u>visibly</u> modified or removed the composition, shape, boundaries or arrangement of the particles composing the rock. In rocks where cementation has been accompanied by solution, recrystallization, dolomitization, or other diagenetic processes, the original sedimentary character may be visible but will be overlain by a secondary crystalline structure. Such rocks are classified as diagenetic.

Rocks which have grown in situ, e.g., certain evaporitic and chemical carbonates such as travertine and tufa, are essentially diagenetic in that the calcareous solutions producing them and the growth locations are both derived from some preexistent carbonate rock. In the marine successions encountered in petroleum exploration, such rocks are rare and of little significance other than as detrital fragments.

It is not necessary to include the terms "sedimentary" and "diagenetic" in a rock description. These aspects of the rock's character should be made obvious from the descriptions themselves. (Similarly, it is not necessary to include the terms "arenaceous" or "resistate" in the description of a rock you have already named a Quartz Sandstone.) However, recognition of this first distinction should influence you in proceeding through the descriptive process.

Dolomites commonly fall within the class of diagenetic carbonates, although rarely; apparently, primary dolomite rocks occur.

NOTE

> Use of the term "dolomite" to describe both the mineral formed from calcium magnesium carbonate and any rock consisting of the mineral as its major constituent can be ambiguous and has in the past lead to confusion. The terms "dolstone" or "dolostone" have been proposed as an alternative and unambiguous name for a rock comprised of dolomite mineral. Unfortunately, these terms have not found wide acceptance and thus are not used in this manual. In this text, at any point where the term dolomite may prove ambiguous, reference to the rock will be made as "dolomitic rock" in order to avoid confusion.

3.30 PARTICLE SIZE

As with detrital sediments, the size of the particles is a major influence on the eventual grain size of a carbonate rock, although in carbonates the diagenetic and other secondary changes to both the matrix and the pore spaces tend to complicate this relationship. Particle size is therefore the next most visible, macroscopically or microscopically, and descriptively the most important characteristic to be included in the rock description. The standard Wentworth size names are used (Figure 3-6) but, unlike any of the classifications in paragraph 3.1, no environment or compositional function is implied in the term. For example, a rock description

NAME	SIZE LIMIT (mm)		COMMENT
	LOWER	UPPER	
Coarse Clastic Carbonate	64.0	—	Rarely an allochemical sediment. Usually terresterial boulder conglomerate derived from a previously consolidated rock.
Calcirudite Dolorudite	2.0	64.0	—
Calcarenite Dolarenite	0.0625	2.0	May be subdivided like sands
Calcisiltite Dolosiltite	0.004	0.0625	—
Calcilutite Dololutite	—	0.004	May be ultrafine grained clastic or orthochemical "micrite"
Very Coarse Crystalline	1.0	—	Diagenetic carbonates only
Coarse Crystalline	0.5	1.0	Diagenetic carbonates only
Medium Crystalline	0.25	0.50	Diagenetic texture or cement
Finely Crystalline	.0625	.250	Diagenetic texture or cement
Microcrystalline	—	.0625	Includes cryptocrystalline, aphanocrystalline

Figure 3-6. Recommended Grain Size Classification

beginning: "Limestone, calcirudite," will be followed by terms detailing the type and composition of the coarse-grained sedimentary limestone that the name implies. Similarly, the term "calcilutite" implies only that the rock is a Limestone, consisting of clay-sized particles with no visible evidence of diagenetic crystallinity. No implication is nor should be made as to whether the fine-grained material is allochemical clastic material or orthochemical micrite. Such distinction is probably not possible, and certainly cannot be made at the wellsite.

Although as all of the classifications suggest there is a definite, often simple relationship between the type of sedimentary environment and the sediment produced, the relationship should not (and indeed in the case of carbonates, often cannot) be implied in the rock name. This system attempts to remove that implication, leaving the geologist to report simple observed fact.

The Wentworth size classification is used for crystalline cement or diagenetic carbonate texture in the same way as in the other classifications (Figure 3-6). Where original sedimentary texture has not been entirely obliterated by recrystallization, report both the diagenetic and the sedimentary texture; for example, "Limestone, medium crystalline, with relic calcarenite texture, ..." Conversely, where recrystallization is only partial, the description would reflect this, e.g., "Limestone, calcarenite, medium crystalline in part, ..."

3.31 PARTICLE TYPE

Where it is possible to determine the origin and history of a carbonate from the shape, size and structure of its particles, this can be most useful both diagnostically and as recognition criteria. However, it is again important to stress the limitations on wellsite work. You must report only what you see!

3.32 Carbonate Minerals

Unlike detrital rocks, carbonates are comparatively uniform in mineralogy. Apart from associated detrital fragments and accessories, three mineral species will be dominant in carbonate rocks throughout the sedimentary section.

3.33 Aragonite: This is a form of calcium carbonate of both organic and inorganic origin. In older rocks, it is much less common than calcite. At normal temperatures and pressures it is metastable and readily inverts to calcite, even in relatively young sediments.

Primary precipitation of calcium carbonate ("whitings") from seawater occurs as aragonite, resulting in aragonite muds and hence aragonite ooliths and pellets. This material readily and rapidly inverts to calcite, and rarely does any aragonite survive early compaction.

Many organisms build shells of aragonite, while in others, such as many lamellibranchs, alternating layers or segregations of calcite and aragonite are found. Aragonite, of organic origin, will gradually invert to calcite but more slowly. On occasion, shells as old as upper Mesozoic are found to contain some aragonite. In general, the mineral is rare, and discrimination of aragonite from calcite is of little interest.

3.34 Calcite: This is the primary constituent of all limestones. Approximately half of all fossil material is primary calcite and the remainder, aragonite, readily inverts to calcite. It is useful to note here that this readiness to invert crystal form will result in fossiliferous rocks, consisting predominantly of aragonite-building organisms, being in general better consolidated and cemented early in the lithification process. Calcite fossils will not recrystallize so early and are more likely to be discovered in an unconsolidated state. Thus it is common to see a lateral or vertical change in induration and reservoir properties of a carbonate rock resulting from faunal population changes.

3.35 Dolomite: This is a mixed carbonate of the form $CaMg(CO_3)_2$ and may be considered as a mixed lattice (c.f. "mixed layer" montmorillonite/illite clays) mineral with alternating "layers" of calcite ($CaCO_3$) and magnesite ($MgCO_3$). Although commonly the composition is fairly close to pure $CaMg(CO_3)_2$, dolomites can contain as much as five mole percent excess calcium, and replacement of magnesium by ferrous iron results in the brownish tinge characteristic of dolomite rocks.

Dolomite rocks are predominately secondary in origin, resulting from the reaction of magnesium compounds with calcite or aragonite. Dolomitization is a selective process, very much dependent upon the temperature and current fabric of the rock. Penecontemporaneous dolomitization, which occurs while the sediment is still unconsolidated, is widespread, uniform and restricted to a single horizon, and probably is a seabed phenomenon resulting from changes in seawater pH shortly after sedimentation. "Subsequent dolomitization" occurs after lithification and is selective, sometimes replacing only matrix and sometimes only shell fragments, depending in part on whether previous recrystallization has occurred and whether aragonite is present. Migration of magnesian fluids is by way of faults and joints in the rock, and dolomitization following joint patterns may be patchy on a regional, or smaller, scale.

A few truly primary dolomite sediments are said to occur, but whether this actually is the case, or (more likely) rapid penecontemperaneous dolomitization, is unsure. Most true primary dolomites are supersaline deposits, and it is common that cyclical shallowing of saline lakes or marine inlets may result in the alternation of primary and secondary dolomites. In a deep marine environment with normal limestone development, shallowing results in increased alkalinity due to shallow saline plant growth or supersalinity due to evaporation. Either case will encourage the precipitation of primary dolomite or, in extreme conditions, pure magnesium carbonate (magnesite). Underlying sediments will undergo secondary dolomitization from the top down.

In addition to the carbonate minerals, most carbonate rocks contain at least trace amounts of similar sized detrital grains, e.g., clays and silts in calcilutites and quartz grains in calcarenites. Due to the deep or sheltered water environment common to much carbonate deposition, such detrital material usually is well sorted and clean.

3.36 Origin

Despite the simplicity of mineralogy, carbonate particle type is extremely variable in form and structure. Particle type will be the primary control of rock texture in a sedimentary carbonate. In a diagenetic carbonate, particle type may be recognized as a relict within the crystalline structure or may be retained strongly and be recognized as having been a control over the pattern and distribution of recrystallization (see Dolomite, paragraph 3.35). For ease of classification, particles may be grouped into two major types, clastic and intact; that is, broken fragments and complete particles as originally formed. This gross classification serves little more than as a basis for further subdivision into the descriptive terms listed in the following paragraphs and shown in Figure 3-7.

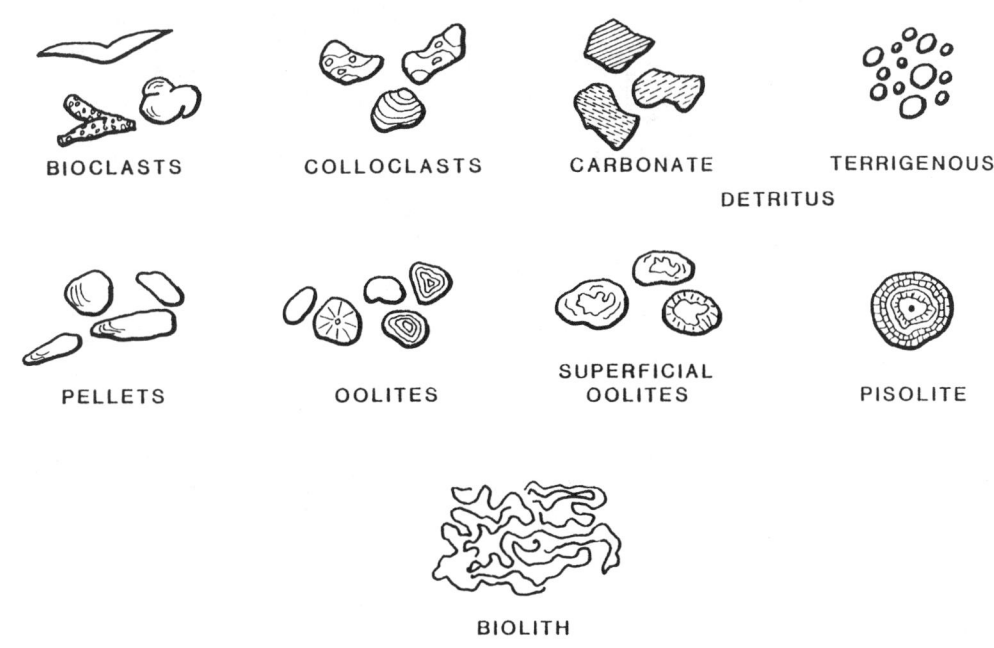

Figure 3-7. Carbonate Particle Type

3.37 Clastic Debris

Many important petroleum reservoirs are formed of clastic sedimentary carbonates. As in detrital rocks, the term clastic has the specific meaning of being a fragment broken from a recognizable whole (parent rock or organism) and transported to its current location.

3.38 Bioclasts: These consist of all debris made up of the supporting or protective structures of plants or animals. Fragments must be identifiably organic and transported. The term is sufficiently elastic to encompass whole tests of small benthonic or planktonic organisms and the remains of benthonic and sessile organisms which may have been transported only small distances on the seabed from their living environment to the point of deposition. Fragments of reef-building growth structures are included in this class if the fragments are identifiably broken and out of place.

It is desirable in describing a bioclast to identify some form of phylum or class since the genetic significance of the presence of differing organisms may be major. Many sessile marine fauna exist only within very strict ecological parameters. The presence of a significant unweathered population of such a fauna gives strong evidence of the presence of such an environment close by the site of deposition. The presence of unattached benthonic and planktonic fauna is of little diagnostic value with respect to environment although the presence of strong weathering on bioclasts is itself indicative of the environment of deposition.

Some calcilutites and calcisiltites, commonly referred to as "chalks," may be of bioclastic character, consisting of the tests of coccoliths or submicroscopic fragments of larger organisms. Since the organic character of such rocks cannot be confirmed at the wellsite, do not use the term bioclast.

3.39 Colloclasts: These are irregular or lobate accretions of calcilute or calcisiltite, commonly of silt or sand-grain size, but occasionally gravel size or larger. They commonly result from the turbation of poorly consolidated seafloor sediments and may have complex inner structures and may show rounding from abrasion and transport (see Intraclasts, paragraph 3.9) although evidence of reworking is not required nor is it always capable of distinction. They are penecontemporaneous with the sediment in which they occur (since major transport would destroy them) and hence are lithologically similar.

Colloclasts may also result from algal action, and this type is marked by irregular concentric rings within the fine sediments. Other origins include concretion around shell fragments or plant roots. In general, colloclasts are distinguished from pellets by having an internal structure and from ooliths by the irregularity of that structure. They also tend to a less regular shape, but this may be lost by reworking. They are indicative of a shallow water reworking environment and possibly one of cyclically varying water depth, such that periods of immersion and aragonite and algal cementation (or conversely, exposure and sun-baking) are alternated with periods of littoral turbulence.

3.40 Carbonate Detritus: This includes particles of all shapes and sizes derived from previously lithified carbonate sediments. Although by definition they are limestones, the criteria for describing such rocks and the environmental significance of their characteristics are similar to those used in assessing clastic detrital rocks.

3.41 Terrigenous Detritus: This includes all noncarbonate detrital material contained in a carbonate rock. Most commonly this consists of clay and silt and sand-sized quartz grains. The amount, grain size, shape and sortings of such detritus may be indicative of the environment of deposition (see Section 2).

3.42 Intact Material

This is material which retains its form and structure from genesis to deposition. Commonly untransported, it shows little or no evidence of physical damage or abrasion prior to deposition.

3.43 Pellets: These are rounded, elongated ellipsoid or discoid sand and silt-sized particles. They are distinguished from colloclasts by their regular outline and total lack of internal structure. They are faecal in origin and therefore any population tends to be of a uniform size and shape which will be characteristic of the organism producing them; and they are lithologically similar to the mud in which the organism lived — lime mud or highly calcitic clay.

Once formed, they have a hardened surface, and if dispersed by current action they would survive some degree of transport. However, substantially transported pellets are difficult to distinguish from other transported carbonate detritus.

3.44 Oolites: These include any spherical or subspherical well-rounded grain made up of regular laminated concentric layers of calcite or with a regular radial structure. True oolites may have a small nucleus consisting of a preexisting sand grain, pellet or shell fragment and do not exceed 2 mm in diameter. For the purpose of this classification, include only the larger concentric particles (pisolites) and those with larger nucleii and only a thin calcite shell (superficial oolites).

Oolites commonly form in shallow water saturated with calcium carbonate. Agitation and heat result in the precipitation of aragonite onto some preexistent sand grain or shell fragment. Where the grain is stationary, radial growth occurs. Concentric growth indicates movement of the particle, probably tidal rolling. Pisolites may have a similar origin but may also form by a mechanism similar to algal colloclasts. Some geologists do not distinguish on the basis of size but on origin, calling all inorganically formed particles oolites, and those with the crinkly lamination indicative of algel accretion, pisolites.

After production, oolites may be transported or reworked. Broken, deformed or weathered oolites will give evidence of extensive transport, and the presence of oolitic concretions may be indicative of reworking after cementation.

3.45 Bioliths: These are commonly the result of reef-building organisms such as red algae or coral bioherms. Molluscan shell banks (coquina) may also constitute a biolith, but this may be difficult to determine from cuttings. Bioliths are characteristic of a shallow, faunally rich environment. By their development they create such an environment and are inconsistent with turbid water that would be associated with detrital rock sedimentation. They will be associated with bioclasts resulting from fragments of the reef builder and other fauna. It may be difficult to discriminate in cuttings between the reef itself and associated fragmentary material. A core will give strong evidence of the massive and undisturbed structure of the biolith.

3.46 PARTICLE SHAPE

The particles contained in a limestone may vary from almost perfectly spherical oolites to extremely angular and elongate shell fragments or totally irregular colloclasts. These shapes are defined by the genesis of the particles themselves. The terminology for roundness and sphericity used for detrital rocks (Figure 2-2) may be used to describe such particles, but it should be remembered that inherent shape will be a control over shape after abrasion and weathering. The two must be considered separately.

An oolite, being well-rounded and spherical in origin, shows little improvement in these qualities even on extensive transport. Thus, rounding is not in this case a sign of transportation. Evidence of transportation in originally rounded particles is difficult to ascertain at best. Evidence from surface texture, the exposure of inner layers at the surface or the presence of broken oolite particles (especially when broken and re-rounded so that the inner structure is exposed on a rounded surface) must be carefully sought.

Shell fragments, being brittle and platy or fibrous in structure, tend to split into smaller fragments upon transportation. Thus fragmentation may be evidence of transportation although scavenging organisms can also fragment bioclasts. Some rounding may occur, but rarely will platy particles become spherical.

3.47 SURFACE TEXTURE

The significance and terminology of carbonate surface textures are the same as used for detrital rocks (paragraph 2.8). The lower hardness of carbonate minerals (calcite: 3; aragonite and dolomite: 3.5 to 4 on the Moh's Scale) compared to quartz (7), and their higher solubility, result in surface texture being less defined and easily lost after deposition.

Surface texture may be the only available evidence of transportation for particles which are rounded in an unabraded formation. Thus, search for evidence of scouring, pitting or breakage.

3.48 SORTING

Similar to detrital rocks (paragraph 2.9), sorting of a carbonate rock's constituent particles is critical in defining its porosity and permeability. This is true even if the rock is recrystallized because, as has been seen above, the distribution of particles, matrix and pore space has some control over the eventual recrystallization distribution.

Sorting as a key to porosity and permeability may be described and interpreted exactly as outlined for detrital rocks in paragraphs 2.9, 2.30 and 2.31. The observations and conclusions will be similar in all essentials.

As with rounding, the implications of sorting will not be so simple when applied to the transport of carbonate rock particles. What appears to be a well-sorted sediment may never have been transported or indeed sorted. The similarity of grain size may be a result of the original particles having been of similar size at the time and place of formation; for example, pellets of a single organism or intact microfossils.

Conversely, what appears to be a poorly sorted sediment having markedly different grain size may in fact be well sorted when mass and shape rather than size are considered. Particles of differing origin may have very differing mass distribution; e.g. chambered organisms, which will affect their densities. Remember (from Section 1) that processes which may appear to sort by size are in fact sorting by mass, cross-sectional area and diameter-to-thickness ratio. When particles of organic origin having radically different structures are considered, it is found that a poorly sorted sediment (that is, containing a diversity of grain sizes) may have been extensively transported.

3.49 FABRIC

Lineation of particles in order to develop a microscopic rock fabric is less important in carbonate rocks due to the absence of sheet-forming minerals like the clays. However, fabrics such as imbricated bivalves may occur. The major influence on the development of carbonate rock texture is recrystallization (discussed in paragraph 3.52.)

3.50 STRUCTURE

Bedding is as important and characteristic in carbonate rocks as in detrital rocks (paragraph 2.11). In cores, the bedding type, regularity and dip should be described along with any characteristic marking of the bedding planes, e.g., scouring, mud cracks, ripple marks, slump structures, burrowing marks or tracks, etc.

In diagenetic carbonates, the process of recrystallization may have removed any direct evidence of bedding. However, they may retain banding, mottling or other rhythmic features in the rock which may be deduced to have been inherited from the original bedding and to now be mimicking it. Carefully describe such features.

The most significant aspects of carbonate rocks are postlithification voids, e.g., fractures, fissures, joints, vugs, caverns, etc. These have major effects in terms of rock strength, porosity and permeability. In oil exploration they are significant both in terms of reservoir potential and lost-circulation problems. Often, carbonate rocks form reservoirs even though the rock itself is too fine-grained to have sufficient effective porosity or permeability to sustain continued flow. Voids and fractures in these rocks provide collection points and permeability enhancement sufficient for the reservoir to produce commercially (e.g., the Ekofisk field, Norway).

The major cause of these voids is solution by moving groundwater. Flow along preexistent fractures or bedding planes results in solution and cavitation. Preferential solution or organic shelter structure may result in cast or moldic porosity. The solution of large caverns or cavities followed by collapse, so-called autobrecciation, is a common feature of limestone successions and, when exposed or reflected at surface, leads to what is termed "Karst topography."

Solution is commonly accompanied by or alternates with re-precipitation and crystallization. Fractures and channels may be filled and sealed by the crystallization of calcite or dolomite. Partial in-filling of channels and fissures may leave isolated voids ranging from small to extremely large. These, along with other isolated nonextensive voids, are given the general title of "vugs" (paragraph 3.59, Porosity).

Fissures, vugs and other nonextensive voids should be described in detail, including shape, size, distribution, infilling, and cause, if these are identifiable. Vugs can be so small as to allow this to be done in cuttings and so large as to be impossible to gain a fair representation even in cores.

Fracturing should be carefully noted, measured and described. Subvertical, slightly curved fractures commonly occur in carbonate cores, especially in more indurated and lower permeability rocks. These result from stresses suffered by the rock during coring and retrieval and should not be confused with those present in the rock in situ. Preexistent fractures are commonly marked by secondary characteristics such as slickensides, staining, secondary crystal growth, or solution channels. By observing such features, genuine fractures may be confirmed. Similarly, by inspecting cuttings carefully, evidence of such features may be found on occasional surfaces, confirming the presence of fractures which may have been suspected from high and irregular drilling torque values.

In cores, the dip and spacing of fractures or joints should be noted and, when there are multiple sets, their relative orientation should be reported.

3.51 MINERALOGY

As stated in paragraph 3.50, the mineralogy of carbonate rocks is comparatively simple, consisting in general of the three major carbonate minerals with minor detrital accessories. However, the instability of one of these (aragonite) and the ready solubility of the others (calcite and dolomite) lead to many complex changes in the distribution and crystal form of these minerals.

3.52 Recrystallization

Inversion of aragonite to calcite is the most ready recrystallization process and is undergone in most carbonate sediments early in the process of sedimentation. Although the process may lead to some bonding at grain boundaries and contribute to the early induration of the rock, little textural change (observable at the wellsite) occurs, and the rock may be considered to retain its true sedimentary character.

Recrystallization, that is, change of crystalline state without change of chemical composition, occurs most readily in calcilutites and in the fine-grain calcilutaceous (lime mud, micrite) matrix of coarser grained carbonate rocks. When occurring in a minor way, this will be recognized at the wellsite by an increased hardness or induration. Since a microfine-grained texture is being replaced by a microcrystalline texture, little or no visible change will be evident.

Major recrystallization to a recognizably diagenetic limestone will again initially and preferentially occur in fine-grained rocks or in the matrix of coarse-grained rocks. Recrystallization commonly appears in patches or along fracture planes and other flowpaths, spreading until conversion to euhedral or anhedral calcite, or spar, has been achieved. During this process, solution of contained coarser particles may occur, leading to the development of moldic porosity. This more commonly occurs with "floating grains" in a predominately fine-grained rock, but has been known to occur in oolitic limestones, producing the subspherical high porosity known as oomoldic.

In coarse-grained sediments, recrystallization of the grains occurs more slowly or later than that of the matrix. This may first result in crystalline casts of the particles, preserving the external details but with total loss of internal structure. With more extreme recrystallization, the whole rock will consist of spar calcite of uniform crystallinity with only faint traces of the original particle outlines.

3.53 Dolomitization

Dolomitization is the most common type of authigenesis (mineral growth within a consolidated rock) occurring in carbonates. Authigenesis is similar to recrystallization, but the process involves a change in chemical composition due to the influx of new material in solution in pore water or the incorporation of material from other components in the rock.

The most likely mechanism for the diagenetic dolomitization of limestones is by an influx of magnesium-rich meteoric water. Such a process would commonly occur shortly after sedimentation with shallowing, e.g., reefs, coastal tidal flats, and sabkhas. Magnesium enrichment occurs by evaporation of marine waters in the supra-tidal regime sufficient to allow the precipitation of gypsum, thus removing calcium and relatively enriching the water in magnesium. The denser magnesium-rich waters flush through previously deposited limestones and dolomitize them. This process, known as seepage or evaporative reflux, has been observed in modern sediments and is believed to be a major origin of secondary microcrystalline dolomites (Figure 3-8).

An alternative mechanism for the (relative) magnesium enrichment of waters in sediments at seabed or only shortly after burial is the removal of calcium due to the calcification of sodium montmorillonites. Such a process would account for the dolomitization of argillaceous carbonates in deep oceanic sediments which could not be the result of evaporative reflux. It is observed that such sediments show increased calcium enrichment of clays in conjunction with dolomitization.

Some dolomites show evidence of having been produced long after sedimentation and compaction. They may be regionally extensive but cut across bedding planes and other depositional features. Such dolomitization occurs in the deep subsurface away from circulating ground water. The source of magnesium enrichment may be due to the leaching or solution of detrital or secondary minerals. This cannot explain more than a few cases, and it must be assumed that increased temperature with burial stimulates dolomitization from normal pore waters without significant magnesium enrichment being necessary.

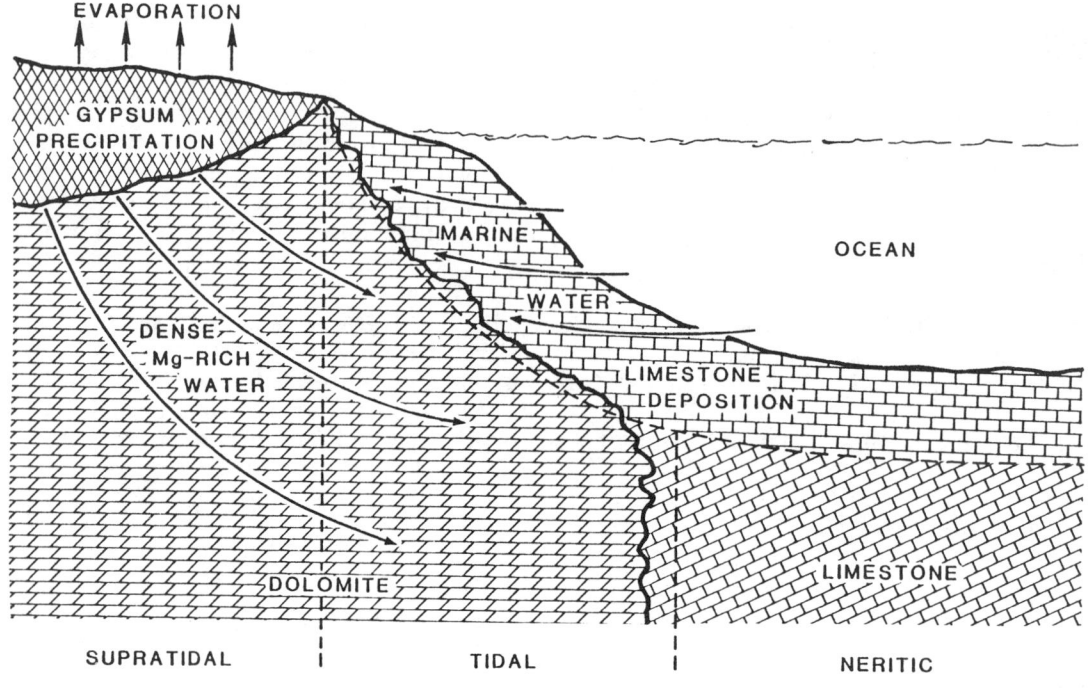

Figure 3-8. Dolomitization by Evaporative Reflux

Dolomite has a strong tendency to develop as single euhedral rhombohedra. This, with the brownish coloration given by iron substitution in the crystal lattice, leads to the loose crystalline structure with excellent intercrystalline porosity, reminiscent of brown sugar, which is termed sucrosic and is so typical of dolomites.

Growth of dolomite within a limestone is selective. Dolomite often will grow in replacement of fine carbonate mud or parts of crystalline grains. Preexistent pore space or vugs may be intruded by crystal terminations, but they will rarely be filled. Thus the original carbonate texture can be retained, though distorted. Organic skeletal remains which resist dolomitization may be dissolved to supply carbonate ions for the dolomitization process, leaving moldic porosity.

Since the dolomite lattice has a 12 to 13 percent smaller molar volume than calcite, and normal pore waters are naturally poor in carbonate ion, a larger volume of calcite must be dissolved than is replaced by dolomite. Dolomitization therefore increases porosity in a closed system. The uniformity and greater size of the dolomite rhombs gives a further increase of porosity and of permeability. The avoidance of dolomite growth in pore space prevents the formation of extensive dolomite cement which, if present, is commonly replaced calcite cement and not secondary. The factors lead to dolomites constituting more favorable reservoirs than limestones.

3.54 Solution

All carbonates are soluble and will eventually dissolve, given a plentiful supply of low saturation pore waters. Factors which affect the selective solution of parts of a carbonate sediment are as follows.

- Mineralogy: Aragonite and dolomite are more soluble than calcite.
- Particle size: Small elongate particles and crystals have high surface/volume ratios and are more susceptible to solution.
- Particle porosity: Some bioclasts are nonporous and resist solution.
- Organic matter: Aerobic decay of organic materials will produce acidic conditions, stimulating solution.
- Proto-petroleum: Anaerobic decay of organic materials produces oily compounds which coat particles, reducing permeability (of the particles) and preventing water contact and solution.

The movement of unsaturated water through a limestone or dolomite (either near to surface or at depth) results in selective solution, governed by the processes above, to form cavities which may range from millimetres to metres in size. Where there is a preexisting major flowpath (e.g., joints or fractures), solution may be less selective — the joints being widened along their length regardless of rock fabric, with locally enlarged cavities where joints intersect. The formation of large cavities may so weaken the rock structure as to cause collapse and the formation of a solution breccia. Alternatively, cavities may contain clastic carbonate particles derived from elsewhere and deposited by flowing ground waters.

For solution to occur, the water must be undersaturated in calcium and/or magnesium. If saturated, precipitation may occur; and secondary crystallization may result in the refilling of voids with crystalline spar calcite or dolomite. (Dolomite growth in voids is less common but can occur.) The presence of coarse euhedral crystals without relict texture is evidence of unimpeded growth in cavities.

Where pore waters are calcium-enriched, e.g., meteoric waters falling onto a gypsum outcrop and percolating into near-surface carbonates, de-dolomitization may occur, resulting in polycrystalline aggregates of calcite mimicking the form of preexistent dolomite rhombs.

In addition to undersaturation, the formation of solution cavities also requires the passage of large volumes of water through the rock. Carbonates are soluble in water, but their solubilities are quite low. For this reason, it is believed that carbonates containing extensive solution-produced porosities have been subject to high-velocity and -volume water flows. The presence of sedimented silt-sized carbonate detritus supports this. Such velocities are uncommon within rocks which are completely filled with water, suggesting that such solution features are indicative of subaerial exposure.

Solution at depth does occur, but it is commonly pressure-motivated and accompanied by recrystallization. This "intrastratal" solution occurs along mobile fronts perpendicular to the direction of applied pressure. Loss of void space within bioclasts and microscopic muds may result in the occurrence of vugular voids, but these will be smaller and less common than the subaerial equivalents. Insoluble organic and detrital material will be carried forward by the hydrodynamic solution fronts. These are commonly seen in cores as stylolites, black waxy or pitch-like, millimetres thick, lobate or sutured linear partings in the rock. With careful inspection, fragments may be recognized in cuttings.

3.55 INDURATION

Induration of carbonate rocks proceeds predominately in the early stages (prior to diagenesis) by compaction and expulsion of fluids. Finer grained calcilutites may show much greater reductions in volume with sedimentary loading than the coarser grained rocks which retain a much larger proportion of their initial porosity prior to cementation.

Calcite has a lower strength than quartz and other detrital minerals. Many organic carbonate particles are complex and contain internal cavities. Compaction often leads to breakage, cracking or distortion of particles in the rock. Such damage may be distinguished from breakage in transport or reworking by the freshness of broken surfaces, close proximity, or even partial attachment of broken fragments of particles.

Descriptive terminology for induration of carbonates is the same as that used for detrital rocks (paragraphs 2.18, 2.19 and 2.20). The clastic and detrital terminology may be used for particulate and crystalline carbonates, and the hydrolysate terminology for lutites.

3.56 CEMENTATION

The differences among secondary mineralization, cementation, and diagenetic changes are strictly of degree. In most carbonate rocks, all three processes result from concurrent solution and crystallization in the rock mass, although they may not be occurring at the same point within the rock.

As stated in paragraph 3.28, some degree of recrystallization may occur within the rock without true diagenesis resulting. The relative severity of the processes, though not necessarily their chronological order, may be described as follows:

- Sedimentary recrystallization: aragonite to calcite inversion, minor recrystallization within the ground mass of insufficient scale or extent to effect particle outlines

- Secondary mineralization: crystal growth on, but not within, particles extending into preexistent voids, living cavities, etc.

- Sedimentary cementation: extensive crystal growth within voids resulting in the closure of interparticle spaces and the bonding of particles

- Diagenetic cementation: loss of particle boundaries due to pressure solution and recrystallization at particle contacts or growth of crystals from voids across particle boundaries

- Diagenetic recrystallization: crystal growth within particles, partial or total loss of particle margins and internal structure, development of secondary voids due to solution
- Diagenetic mineralization: secondary crystal growth within diagenetic voids

It can be seen that this process may be continuous and the rock may undergo numerous cycles of diagenetic change. Fracturing, dolomitization and de-dolomitization may further complicate and influence the process.

The geologist must endeavor to identify and describe the individual components of the rock and its (implied) history. Thus, in describing a limestone, you may, for example, itemize the rock character as follows:

1. Crystalline limestone; size and form of crystals.
2. Relict sedimentary texture; particles, matrix, and original cement.
3. Secondary voids; size and distribution.
4. Fractures and solution features.
5. Dolomitization.
6. Secondary mineralization in diagenetic voids, subsequent to recrystallization.

Such detail is essential in establishing the true nature of the rock, both as a potential reservoir and as evidence of its history.

Where a rock consists of essentially unaltered calcarenite or larger particles in a crystalline groundmass (or even when identifiable relics of such are visible), it is important to recognize whether this is spar cement (sedimentary cementation) or preferentially recrystallized calcilutite matrix (diagenetic recrystallization). In terms of Folk's classification (paragraph 3.15), this means a decision as to whether the rock is a sparite or a recrystallized micrite.

Probably the simplest rule in making this discrimination is Folk's own definition (paragraph 3.16). Lutite matrix is an integral part of the deposited sediment, whereas spar cement forms after deposition, growing within a framework defined by the particle packing present. (However, it may occur very soon after deposition; glass bottles and fishing debris are found in well-cemented calcarenite "beach rocks" less than 20 years old. Thus any rock in which interparticle contact does not occur (the particles are "floating") must have originated as a micrite sediment which included coarse particles. A rock with good intergranular contact, that is, "grain-supported," may be a cemented calcarenite — but this is not conclusive. Further evidence is required to distinguish recrystallized matrix from secondary cement in a true grain-supported calcarenite or calcirudite.

The essential difference between the two forms is that cementation is the result of crystallization from an aqueous solution with unimpeded growth into a void. Conversely, matrix recrystallization occurs at the crystal lattice level in the solid phase. (Microvoids do occur within calcilutites, and it has been proposed that growth of a crystal within such a void occurs from material derived from solution of matrix adjacent to the void, thus enlarging it or producing a new void. Although nucleation (formation of the initial new crystal) may occur this way, it is probable that further growth occurs by reordering at the lattice level within the solid phase.)

Even when the original void outline is lost, crystal form will be indicative of its presence. Crystals will be euhedral, of uniformly increasing size with distance from the void wall and with long axes perpendicular to the wall. Crystals will grow in optical continuity except at the wall where they will be abruptly terminated and at the center of the void where growths from each wall meet with no particular crystallographic orientation, so-called "compromise boundaries" (Figure 3-9).

Figure 3-9. Spar Cement and Recrystallized Matrix

Growth in a matrix may begin at isolated points and proceed at varying rates (paragraph 3.52). A recrystalized matrix will have a fabric in which little ordering of crystal size or orientation will be present. Crystal boundaries will be compromised in the case of coarse crystals. Fine crystal growth and form will be defined by the faces of preexistent adjoining coarse crystals. An intimately bound texture similar to some igneous rocks may result (again, see Figure 3-9). Enclosed voids (e.g., chambered organisms) may have been preserved from matrix fill and may therefore remain empty or contain uniformly crystalline spar, as above.

When using Folk's classification, such distinction must always be made since, without it, a rock term cannot be selected. Even when using another (or no) standard classification, the distinction should be attempted and included in the description. An experienced geologist should be capable of recognizing the distinctive features even when further recrystallization results in a secondary crystalline fabric overlaying an original crystalline fabric containing relict sedimentary features.

Secondary mineralization, sedimentary and diagenetic cementation will all contribute to the induration and strength of the rock. Nevertheless, so many other factors influence these that it is not recommended that such terms as "weakly" or "strongly" be used. Preferable terms include:

- partially
- poorly
- moderately
- well
- very well
- extremely well

These terms are based upon an assessment of the amount of intergranular cement and of pressure recrystallization at grain boundaries. If the rock is diagenetic, that is, if crystallization across grain boundaries has occurred, resulting in a total crystalline texture, the term "cement" should not be used other than in describing relict texture (if visible).

As stated previously, cementation may occur very early in the process of sedimentation, even in a subaerial or beach environment. In such environments the carbonate deposited will be one typical under those conditions. Thus in seawater this will be aragonite; and where evaporation causes enrichment (in beach rock), high magnesium calcite may be formed. Such cements are seen in recent and Pleistocene carbonates; but in older rocks, conversion to low magnesium calcite occurs. This is usually without any visible change in crystallinity — but on occasion the aragonite crystals may be preferentially dissolved, leaving vugs. These may later be partially or totally filled with calcite, resulting in a texture exhibiting primary and secondary cavity fill.

Recrystallization of cement or secondary crystals may occur without recrystallization of the whole rock, leading to a secondary cement texture. After crystallization of spar cement of uniform texture (Figure 3-9), which is commonly referred to as drusy or drusy mosaic, further growth may occur with smaller crystals growing at the expense of large. The eventual texture, referred to as blocky, is granular in appearance, crystal axes and faces are less well defined, and crystals tend to be equidimensional in all directions (Figure 3-10). Although less distinctive than drusy, the uniformity of size and texture of blocky cement allows it to be distinguished from recrystallized matrix.

DRUSY MOSAIC
(SPAR CEMENT)

BLOCKY TEXTURE
(RECRYSTALLIZED CEMENT)

EPITAXIAL GROWTH
(RIM CEMENT)

Figure 3-10. Drusy, Blocky and Rim Cement in Carbonates

A third type of cement texture may be observed when the particles of which the rock consists are single crystals of calcite, e.g., crinoid particles. In this case, calcite is deposited onto the grain and growth occurs in optical continuity as a single crystal. The eventual texture is one of a coarse crystalline rock with compromise boundaries where crystals meet. There is no parallelism of crystal axes since the orientation of each crystal is controlled by the orientation of the single crystal particle from which it grew. This type of cemented rock, called "rim cement" or epitaxial growth, may be confused with a recrystallized diagenetic carbonate. It may be distinguished from such rocks by its lack of crystal orientation and faces and by the inclusion within each crystal of an unaltered (usually darker colored) sedimentary particle (Figure 3-10). Secondary drusy or blocky calcite may sometimes be seen between epitaxial growths filling intercrystalline voids.

Carbonate rocks may occasionally be cemented by other minerals — for example, anhydrite, silica, siderite, iron oxides, and sulfides and barite. For identification of these minerals, see Figure 2-7.

The most common and abundant noncarbonate cement in carbonate rocks is silica. The process of silicification involves the replacement of calcite by silica, either in an amorphous or crystalline form. This is discussed below (paragraph 3.57, Accessories). The resultant rock is extremely hard and well indurated, often with a conchoidal fracture.

3.57 ACCESSORIES

Minor accessories in carbonate rocks are commonly detrital or diagenetic products of terrigenous rock fragments contained in the original sediment, with some mixed carbonate/terrigenous diagenetic minerals. These are included in Figure 2-7.

The presence of elemental sulfur and metallic sulfide as concretions in carbonates or as staining on joint and fracture planes is significant in indicating anaerobic decomposition of organic material. This may indicate potential hydrocarbons and/or hydrogen sulfide hazards

Accessory silica is chalcedony, chert and crystalline quartz grown in the rock during diagenesis and not originally present in the original sediment as resistate grains. Solution of quartz grains as a source for silicification is only minor. The major source of silica is the decomposition of siliceous organisms, e.g., radiolaria. The shells of these organisms are unstable after death and, with decomposition of the organic tissues, opaline silica is liberated and readily dissolved. Silica-cemented carbonates often contain nodules and vugular fill of structureless chert and silica-replaced fossils. The degree of silicification and nodule formation is directly proportional to the volume of available silica and hence the original population of siliceous organisms.

Although silica cementation may simply be the crystallization of quartz in preexistent voids in the carbonate, very often simultaneous solution of calcite and precipitation of silica occurs. The former process explains the quartz cementation of calcarenites and calcilutites and the filling of vugs to produce chert nodules. The second process must be proposed to explain the delicate and precise replacement of carbonate shells and skeletal remains with microcrystalline quartz. Such replacement is often so precise it defies visual discrimination even in thin section.

Fossils, in addition to being the major component of carbonate rocks, are often their most significant accessories. In addition to the major rock-building organism, the presence of other minor organisms may be significant in interpreting the environment of deposition or source of the sediment (if the two are not the same). The Energy Index Classification of Plumley, Risley, Graves and Kaley (Figure 3-5) is useful when reviewing both rock types and faunal assemblages. Useful information may be gathered at the wellsite without a major expenditure of time or extensive knowledge of micropaleontology.

The presence and abundance of all observable fossils should be noted with identification of at least phylum or class. Note breakage and abrasion of fossils; and for biolithic rocks, attempt to decide whether the organism is in its growth position or if transported fragments are being seen (this will of course be easier in cores). Finally, describe trace fossils, burrows, trails, etc.

3.58 COLOR

Color itself may be of less significance in carbonates than in detrital rocks and accessory minerals. It is, however, most important that color be accurately described. This is because color variance is often so slight in carbonates that recognition and discrimination may depend upon a precise description of shade and hue. A color chart may be useful in this.

Variation in color of carbonates may result from the presence of detrital material or from the substitution of other metallic ions into the mineral lattice, e.g. manganese (pink to rose red) or iron (medium to dark brown).

3.59 PETROLEUM SIGNIFICANCE

Due to their varying composition, carbonates may occur as source rocks, reservoir rocks and cap rocks, but they are most significant as reservoir rocks. Unlike detrital rocks, major changes occur in reservoir characteristics, porosity and permeability throughout the history of a carbonate. Initially, extremely high porosities and permeabilities at sedimentation may be almost totally lost early after burial, only to be regained during diagenesis. The importance of large-scale or irregularly distributed features such as fractures, solution channels and vugs to the reservoir properties of a carbonate make accurate estimates of overall porosity and permeability difficult in cuttings and often even in cores (Figure 3-11).

3.60 Porosity

Pore systems in carbonates are generally complex in their geometry and genesis. Carbonate porosity is polygenetic both in the sense of mode and of time of origin. Porosity can form by the inclusion of voids within the sediment particles, from sediment packing or sediment shrinkage, by the fracturing or brecciation of the rock, by selective solution of particles within the rock, or by the indiscriminate solution of a mass of rock. Pore size may vary from 1 micron (.001 mm) to hundreds of metres in size. (Adams and Frenzel, 1950, classified the "Big Room" at

CHARACTERISTIC	DETRITAL	CARBONATE
Amount of initial porosity in sediments	25 to 40%	40 to 70%
Amount of ultimate porosity in rocks	Commonly half or more of initial porosity; 15 to 30% is common	Commonly none, or only small fraction of initial porosity; 5 to 15% common in reservoir facie
Type of initial porosity	Almost exclusively primary interparticle	Widely varied because of postdepositional modifications
Size of pores	Diameter and throat sizes closely related to sedimentary particle size and sorting	Diameter and throat sizes usually show little relation to sedimentary particle size or sorting
Shape of pores	Strong dependence on particle shape — a "negative" of particles	Greatly varied; ranges from strongly dependent "positive" or "negative" of particles to form completely independent of shapes of depositional or diagenetic components
Uniformity of size, shape and distribution	Commonly fairly uniform within homogeneous body	Variable; ranges from fairly uniform to extremely heterogeneous, even within body made up of single rock type
Influence of diagenesis	Minor: usually minor reduction of primary porosity by compaction and cementation	Major; can create, obliterate, or completely modify porosity; cementation and solution important
Influence of fracturing	Generally not of major importance in reservoir properties	Of major importance in reservoir properties, if present
Visual evaluation of porosity and permeability	Semiquantitative visual estimates relatively easy and accurate	Variable; semiquantitative visual estimate ranges from easy to virtually impossible; instrument measurements of porosity, permeability and capillary pressure commonly needed
Adequacy of core analysis for reservoir evaluation	Core plugs of 1-inch diameter commonly adequate for primary porosity	Core plugs commonly inadequate; even whole core (3-inch diameter) may be inadequate for large voids and fractures
Permeability and porosity relationship	Relatively consistent; usually dependent on particle size and sorting	Greatly varied; commonly independent of particle size and sorting

Figure 3-11. Comparison of Porosity in Detrital and Carbonate Rocks

Carlsbad Caverns, New Mexico, as a "macro pore.") Pores of radically variant shape, size and history may occur within a single rock sample. For this reason a system of porosity classification is required in order to allow disciplined description of porosity in carbonates.

The simplest and most common classification of porosity is as either primary or secondary. If the meanings are clearly defined and carefully used, the terms are descriptively useful, carry no interpretive or genetic implications, and yet convey a great deal of information regarding the reservoir characteristics of the formation. There are also certain pragmatic advantages in their use as shown in paragraphs 3.61 and 3.62.

3.61 Primary Porosity: This is porosity which forms an integral part of the rock fabric. Thus it is <u>primary</u> to the rock; any sample taken of the rock regardless of size (cutting, core, exposure) would display porosity equally. Porosity, in order to be primary, must therefore be of small size and evenly distributed. Interparticle porosity, between sedimentary grains or diagenetic crystals, is the most common form of primary porosity, and voids within skeletal particles and growth structures will commonly be primary. Vugular porosity will often be too unevenly distributed to be considered primary, but smaller scale microvugs (so-called pinpoint porosity) may be sufficiently so to be primary.

Primary porosity can be readily and accurately identified in cuttings (often with the naked eye), measured in core analysis, and will be accurately reported by sonic, neutron and formation density logs.

3.62 Secondary Porosity: This is porosity that is secondary to the rock fabric (although in origin it may have been defined or influenced by that fabric). It is commonly too large in scale or irregular in distribution to be estimated, or sometimes even observed, from cuttings. Even cores may be insufficient for the full extent of the porosity to be seen. Fractures, fissures and vugs are the most common types of secondary porosity, but sedimentary or growth structures may occasionally produce enlarged or irregular secondary porosity.

Where secondary porosity cannot be estimated in cuttings, it may be identified by evidence of preexistent fracture surfaces or void walls on the surface of cuttings, e.g., concavity, solution features, staining, mineral growth or internal sediments (see paragraph 3.54). Additional evidence may be given by irregular rate of penetration, sharp drilling torque increases and decreases, and partial or total loss of drilling fluid. Core plugs provide insufficient volume of sample for reliable determination of secondary porosity, and when fractures, fissures or large cavities occur, even whole core may be unreliable. Sonic, neutron and formation density logs will report unrepresentative or even incorrect porosities where secondary porosity is major.

The advantage of this definition of primary and secondary porosity is that questions of the genesis and history of the rock need not be considered prior to choosing a classification. This is valuable at the wellsite where sample quality, available tools and time may prevent such questions from being conclusively answered. A second advantage as shown above is the pragmatic one that, regardless of origin, primary and secondary porosity have essentially different characteristics both in terms of "measurability" and reservoir behavior. For example, a sucrosic dolomite and a clean calcarenite will have similar primary porosity, permeability relationships and reservoir "deliverability" despite the fact that one is sedimentary and the other diagenetic. Similarly, fracturing of a limestone may be followed by water flow and solution to form fissures and cavities, collapse and autobrecciation. Transport and redeposition of the brecciated fragments may result in detrital coarse clastic carbonate. Throughout the process the primary porosity of the parent rock will remain consistent, and though the secondary porosity may be improved over the process, there will be a strong genetic relationship between the reservoir capabilities of each. It is also worth noting that from wellsite evidence it would be extremely difficult to distinguish the fractured, fissured, collapsed and breccia limestones from each other if they were drilled.

Choquette and Pray (1970) suggested a redefinition of the terms in order to include genetic specificity: "primary porosity either originates at time of deposition or was present in particles before their final deposition. Secondary porosity originates after final deposition." While this statement is generally true of primary and secondary porosity as previously defined, there are important exceptions and, as important, cases where ambiguity exists — preventing a definite conclusion (see the examples above). For this reason, it is recommended that the more general definition of primary and secondary porosity as defined above be retained.

When not recommended for Log descriptions, the system of Choquette and Pary is of value to the logging geologist or wellsite in preparing more comprehensive reports where interpretation and conclusions may be required (and possibly with hindsight). To avoid confusion and to give compatibility with the rock classification used in this manual, the terms Sedimentary and Diagenetic Porosity should be used only when genetic significance is attached to porosity classification (Figure 3-12).

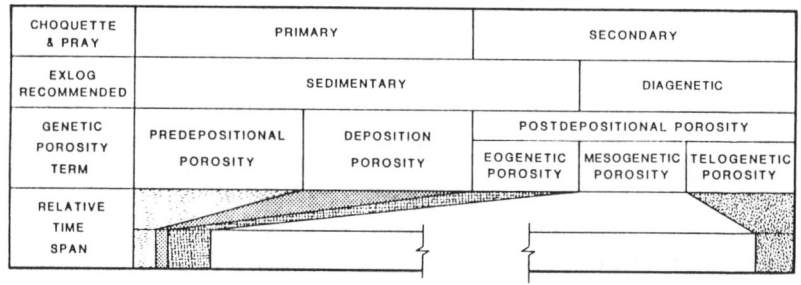

Figure 3-12. Genetic-Porosity Terms and Types

The system attempts to isolate porosity genesis to five distinct stages in the history of a carbonate. These stages may be sequentially arranged and provided with typical time scales for their operation, but no strict chronology is or should be implied in their use. In attempting to combine geometry and genesis of porosity,

Choquette and Pray made the major division (their usage of primary and secondary) between the porosity developed up to and at the point of deposition and that developed afterwards. If the terms primary and secondary are reserved for pore geometry only, and the genetic terms already used for rock development are used similarly for pore development (see paragraphs 3.28, 3.29 and 3.56), we may divide as follows: Sedimentary Porosity (present in the original particle) of the depositional environment; and Diagenetic Porosity, created by influences and processes unrelated to the original depositional circumstances (although, perhaps, historically derived from them). The genetic porosity terms defined by Choquette and Pray are the most important part of the classification and are shown in Figure 3-12 and defined in paragraphs 3.63 through 3.67.

3.63 Predepositional Porosity: This may be formed at any time prior to the final deposition and burial of the sediment. Most typically it will consist of organically derived porosity either of a skeletal, chambered or biolithic origin. Nonskeletal grains such as pellets and oolites may contain appreciable porosity of an intergranular or intercrystalline type, and the intercrystalline type may also be present within the apparently solid parts of skeletal or biolithic remains. Preexistent primary and secondary porosity in carbonate or terrigenous detrital fragments may also be considered to be predepositional to the rock in which they are incorporated. Structures at the sedimentary interface caused by wave action or bioturbation, which rework the bottom but do not significantly penetrate it, may also be considered to be predepositional.

3.64 Depositional Porosity: This is formed by the final interrelation of particles and structures in the rock prior to burial. It is almost entirely interparticle in type since other porosity-developing or -destroying techniques occurring in the depositional environment can be more readily classified as predepositional (e.g. reworking, superficial bioturbation), or immediately postdepositional (e.g. burrowing, mud cracks, organic decomposition). Depositional porosity may constitute up to one-third of bulk volume in calcarenites and two-thirds or more in calcilutites. This porosity is of only short duration, being rapidly lost by compaction and diagenetic processes.

3.65 Eogenetic Porosity: This is formed after deposition but by processes which are characteristic of the environment of deposition and governed by changes and occurrences in that environment. Thus, though not depositional in nature, such porosity and its creation can be said to be part of the sedimentary process. Eogenetic processes commonly occur to only a few metres below the sedimentary interface. However, where eogenetic processes occur due to the circulation of ground waters (for example, penecontemporaneous dolomitization), the results may be seen to depths of hundreds of metres if permeability and hydrodynamics are suitable, thus drawing into question whether eogenetic porosity can truly be said to be sedimentary. (Choquette and Pray suggested resolving the question by letting the eogenetic stage be transitional between sedimentary and diagenetic — as indeed, the two processes are.)

A major source of eogenetic porosity is the penetration by or decay of organic structures which do not fossilize. Burrowing organisms may produce extensive burrow systems which may not be filled, or may be filled with some later soluble material. Similarly, penetration by plant roots will produce extensive porosity after death and decomposition of the plant. Decomposition of plant debris in a buried, unconsolidated sediment will evolve gas. If insufficient sediment permeability exists, this will lead to distension, forming fenestral porosity.

An alternative mechanism for the formation of fenestral porosity is sediment shrinkage due to dehydration when exposed in intertidal or supratidal environments. At surface this will produce mud cracks; below the surface, separation of sediment laminae will produce wide, flattened fenestral pores. (Separation has been said to occur commonly at surfaces formed by algal mats. This may be true.)

The major processes active at the eogenetic stage are not porosity creating but porosity destroying. Porosity is lost by compaction and dehydration in the main, but cementation, recrystallization and dolomitization may also contribute.

3.66 Mesogenetic Porosity: This is created at depths below the major influence of processes directly operating from or related to the surface. Cementation with or without minor solution is the major process active at this stage, and it is where true lithification of the rock occurs. Marked physical compaction and pressure solution occur at the higher overburden pressures of the mesogenetic stage. All six stages of recrystallization listed in paragraph 3.56 occur in the mesogenetic stage and, if sufficiently well developed and identifiable in a succession, may provide a basis for subdivision of the mesogenetic stage.

3.67 Telogenetic Porosity: This is formed only when a previously compacted and lithified carbonate is uplifted and exposed to subaerial or shallow subsurface processes. The term applies only to the interaction of surface processes with old rocks; that is, if the rocks are not totally destroyed by the processes but were subjected to later reburial (with the formation of unconformities), as opposed to the erosion of newly deposited sediments during temporary interruptions in sedimentation, or lacunas.

The thickness of the telogenetic zone is extremely variable, commonly a few metres or less but as much as several hundred metres in strongly uplifted Karst topography. Solution by meteoric water (rain/surface derived) is the most important process in the telogenetic stage, and for solution to have an appreciable effect, vast quantities of water must pass through the rock (paragraph 3.54). Thus paleoclimate, uplift, location of the water table, and initial permeability all will affect the development of telogenetic porosity. The lower limit is commonly gradational with telogenetic solution effects becoming less and less marked. Sometimes the existing water table (if constantly maintained) can serve as an effective lower limit.

Although solution due to large movement of meteoric water is the major process active in the telogenetic stage, other associated porosity-producing and -destroying processes may leave evidence in the rocks (commonly only in cores) of the creating environment. Fracture and brecciation commonly occur in such environments, and evidence may be seen of rock (as opposed to sediment) burrowing organisms. Secondary mineral growth and cryptocrystalline precipitation along and within fissures and internal sedimentation act as porosity-reducing factors in this environment. Chemical effects of the circulating waters may include de-dolomitization and the precipitation of amorphous silica in voids. In general, telogenetic porosity is secondary in nature and, while difficult to identify in cuttings and even in cores, will have major influence over the productivity of a reservoir.

Figure 3-13 illustrates the environments typical of the various porosity genetic stages and the types of porosity typical of them.

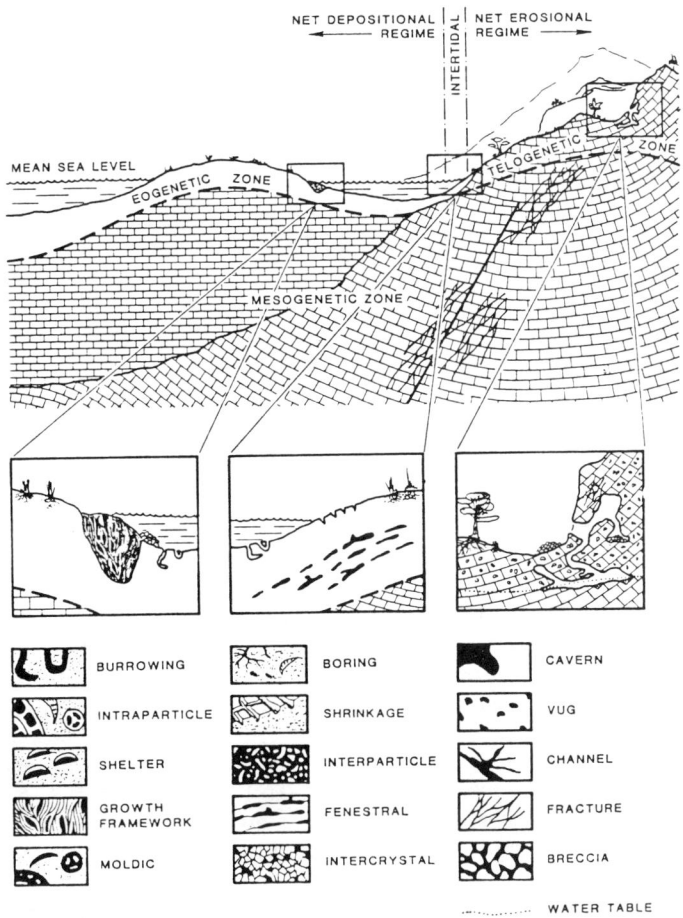

Figure 3-13. Porosity Types and Environments

3.68 Porosity Description

As shown in Figure 3-13, there are fifteen recognizable types of carbonate porosity. These may be grouped according to the environment in which they are commonly created or according to the degree of rock fabric control over their creation, as follows:

- Fabric controlled
  - Interparticle
  - Intraparticle
  - Intercrystal
  - Moldic
  - Shelter
  - Growth framework
- Fabric influenced
  - Fenestral
  - Boring
  - Burrowing
  - Shrinkage
  - Breccia
- Fabric independent
  - Fracture
  - Channel
  - Vug
  - Cavern

Of the fifteen, only seven are significant in the creation of the volumetrically important porosity required for a petroleum reservoir. These are interparticle, intraparticle, intercrystal, moldic, fenestral, fracture, and vugular porosity. The others may occasionally be economically significant and will always be important in rock descriptions both for characterization and for recognition of environment and history. "Compound" pore systems commonly exist composed of one or more of the major porosity types with contribution from the less abundant types. "Gradational" pore systems are sometimes seen where individual pores or groups of pores are intermediate in character between two porosity types or are so interconnected as to prevent separate recognition. Examples of this may occur in the interconnection of inter- and intraparticle porosity, or in the partial solution of moldic porosity margins, resulting in pores which are transitional between molds and vugs.

Porosity is so variable in carbonates that the logging geologist must give as much attention and detailed description to the porosity as to the rock itself. Figure 3-14 gives a recommended sequence for description of porosity which should include abundance, modification process and direction, time of formation, size, and type. All of these may of course be compound or gradational as described above.

3.69 Permeability

The complex interrelationship between particles and pore space in carbonates precludes drawing simple conclusions about permeability from a visual inspection of the rock. In fact, experiments have shown that, while measured porosity in carbonates shows a strong correlation with the fluid recovery efficiency of a reservoir, there is no such correlation between recovery efficiency and measured permeability.

Even sucrosic dolomites, which in general have the best carbonate permeabilities and appear to be analogous in pore structure to detrital rocks, have markedly different fluid flow characteristics from their arenaceous equivalents. For example, few quartz siltstones are productive reservoirs, and those which are have grain sizes at the upper end of the grain size range (.0625 mm). Conversely, there are many dolomite reservoirs producing from interparticle porosity with grain sizes below this, ranging down to 0.01 mm.

Obviously, visual permeability estimates in carbonates are difficult to make, unreliable and of little value in determining reservoir performance. For these reasons they should not be attempted.

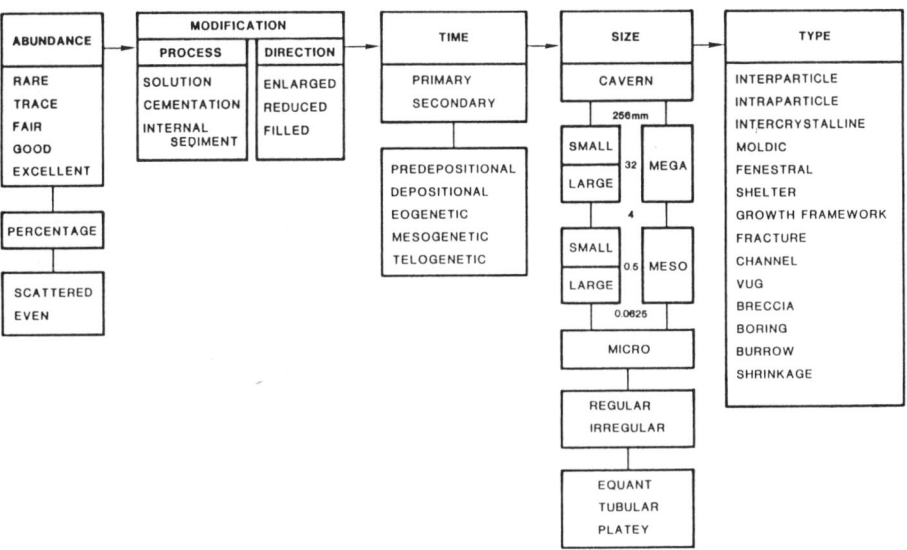

Figure 3-14. Porosity Description (after Choquette & Pray)

4
CHEMICAL ROCKS

4.1 INTRODUCTION

The detrital arenites and argillites and the marine carbonates constitute the vast majority of sedimentary rocks encountered in petroleum exploration. Although of lesser quantitative significance, the rock types discussed in this section have geological importance as markers, environmental indicators, and even, on rare occasions, as source, reservoir or cap rocks.

4.2 NON-MARINE CARBONATES

Most significant occurrences of limestone and dolomite are of marine origin. However, limited occurrences of fresh-water carbonates are seen and may be of diagnostic significance. Fresh-water organic limestones are of minor importance in both geological and petroleum aspects. Most nonmarine limestones result from chemical precipitation and are formed in association with zones of extensive subsurface solution, e.g. Karst processes. Along with other signs of telogenetic processes they are symptomatic of exposure and unconformity.

4.3 PRECIPITATION

Groundwaters percolating through the telogenetic zone rapidly become saturated with calcium bicarbonate which, given a change in environment, readily loses carbon dioxide and precipitates calcium carbonate (e.g., stalactites and stalagmites). Where surface springs issue from limestone, encrustations of cryptocrystalline calcium carbonate, tufa, or travertine may be formed. These are porous and cellular in structure and often banded with other minerals.

The encrustation of calcium carbonate onto preexisting particles in stream beds may result in the formation of inorganic pisolites. These commonly have a radial structure, indicating static growth (see paragraph 3.44) consisting of coarse, fibrous calcite. Occasionally, concentric bands of other minerals, e.g. gypsum, may be present within these inorganic pisolites. The mechanism producing the precipitation of calcium carbonate may be due to a fall in temperature of the water, as this is probably true of hot springs.

An alternative explanation for the precipitation in subsurface caves is the evaporation of water. This appears to be unlikely since modern caves exhibit high relative humidity which would inhibit evaporation, and no increase in ionic concentration of contained water is seen relative to associated ground waters, which would be the case for all ions other than carbonate and calcium if evaporation were taking place. A more likely explanation is that water percolation through the vadose zone (aerated upper soil) becomes supersaturated with carbon dioxide in this generally acidic environment. When entering a cave or the open atmosphere, the carbon dioxide will be lost and calcium carbonate precipitated.

4.4 LACUSTRINE DEPOSITS

Nonmarine environments such as fresh-water lakes support a large though not diverse calcium-carbonate-producing flora and fauna. Similarly, many of the sedimentary processes taking place in the marine environment, both mechanical and chemical, are also present in the lacustrine environment. Thus a range of fresh-water limestones is observed analogous to those of marine origin.

Due to the lower energy and limited areal extent of a lake, these fresh-water limestones tend to be less well washed and homogeneous than their marine equivalents. The admixture of clay material to produce so-called "Marls" and recognition of the limited fresh-water fauna is a common indication of the type.

4.5 SILICEOUS ROCKS

In addition to nodules, concretions and fossil replacement by chert in limestones, as discussed in paragraph 3.57, nonclastic siliceous rocks may also occur independent of limestones. Such siliceous rocks have either a chemical or biological origin.

4.6 ORGANIC

Organic siliceous rocks occur both in fresh-water deposits and deep ocean basins. Such rocks are produced, in the same manner as microfossiliferous limestones, from the tests of planktonic fauna — the most common organisms being diatoms and radiolaria with some contributions from siliceous sponge spicules.

Such organisms are observed to congregate in areas of higher silica concentrations in seawater (e.g. in areas of vulcanicity), at river estuaries or immediately seaward of them. (Fresh water tends to be richer in silica than seawater.) Nevertheless, even in these enriched environments, the water is in general undersaturated in silica due to the resistate nature of detrital quartz. Silica concentration is too low for direct precipitation of silica to occur. Certain diatoms have been observed to be capable of abstracting silica directly from kaolinite, but this appears to be rare. In general, the organisms seem to be capable of a catalytic precipitation of molecular silica in a highly efficient manner.

Silica molecules are adsorbed onto an organic substrate and polymerized to form opal (amorphous silica). Unlike calcium-carbonate-using organisms, the opal test or shell is not a permanent structure, but one which requires continual maintenance to prevent dissolution in the undersaturated water. Upon death, the organism settles to the sea or lake bed and decay of the organism proceeds, releasing the opaline silica as a siliceous ooze. This ooze would slowly dissolve in warm surface waters, but at the low temperatures seen at depth, solution is so slow as to be negligible.

After burial, the opal recrystallizes to crypto- or microcrystalline chert. This recrystallization process is a typical one of simultaneous solution and precipitation. However, circulating ground waters are not necessarily involved in the process since the original water of the opaline silica colloid is sufficient to support the reaction at subsurface temperatures.

Bedded cherts are usually even-bedded, thinly laminated to massive and of the order of 10 feet thick although they may be as much as a hundred feet. They are commonly black or dark gray in color but occasionally mid- to light gray or with a brown or green tinge. Indicative of the environment of deposition, diatomaceous and radiolarian cherts are often rich in clay impurities, while spiculiferous cherts commonly contain unusually large amounts of calcite.

In drilling, the most easily recognized characteristic of cherts is their tremendous hardness (seven on Moh's Scale) and glass-like brittleness. Drilling into a chert with a milled-tooth bit will result in an almost complete halt in penetration, probably bit-bouncing and vibration. Torque will probably be low since the bit teeth cannot penetrate the hard chert surface; but if the chert is thinly bedded or fractured, torque may become irregular as the bit "sticks and skips" over the broken chert surface.

Chert cuttings will be large, elongate, and shell- or blade-shaped with fresh, curved, conchoidal fracture surfaces. Abundant metal flakes and shavings in the sample returns confirm the hardness and sharpness of the formation surface and cuttings.

4.7   INORGANIC

Bedded cherts of an inorganic origin are suspected to have resulted from the direct precipitation of amorphous silica in both shallow marine and lacustrine environments. The mechanism is seen to occur in modern lake sediments, and in ancient marine sediments evidence is given by slump structures and bed deformation, indicating a semisolid silica gel rather than the fluid colloidal ooze deposited in organic sedimentation.

In evanescent seasonal lakes, rapid algal growth during periods of water influx leads to highly alkaline conditions, allowing the corrosion and solution of large amounts of silica from detrital quartz and clay minerals. Decrease in pH due to algal reductions in later seasons or concentration due to evaporation of the lake causes a colloid of amorphous silica to form which will "stiffen" to form a thick viscous gel. A similar supersaturation of silica may result from the devitrification of volcanic glass in areas of intense vulcanicity. In this case, both inorganic precipitation and a large increase in diatom and radiolaria populations may combine to produce thick sediments.

Ancient inorganic cherts are not visually discernible from organic cherts, and no attempt should be made to do so at the wellsite. In younger sediments, ghosts of organic structure may remain in the rock, but in pre-Tertiary cherts these will largely have been lost through recrystallization.

4.8   FERRUGINOUS ROCKS

Iron is the most abundant element in the earth and the fourth most abundant in the crust (after oxygen, silicon and aluminum). Most sedimentary rocks contain iron as a cementing or accessory mineral (for example, red or green sandstones and shales), or substituted within the rock-forming mineral (e.g. dolomite). Many sedimentary rocks, however, are notably enriched with iron, commonly in the form of carbonates, hydrated silicates, oxides, hydroxides, and sulfides. The characteristics of these iron minerals are summarized in Figure 2-10. Such rocks contain more than 15 percent iron while some, the Precambrian "iron formations," can contain as much as 40 percent. Although modern iron-enriched sediments do exist, none have this degree of iron enrichment — probably due to the much higher oxygen content of the atmosphere following the appearance of the earliest photosynthesizers in the late Pre-Cambrian.

4.9   CARBONATES

Siderite (ferrous carbonate) is one of the main constituents of sedimentary ironstones. It forms primary deposits of muds and concretionary nodules. It also occurs as cements, secondary mineralization, and as diagenetic accretions. (See Figure 2-10 for descriptions of minerals discussed in paragraphs 4.9 through 4.12.)

The primary rocks are dense, compact and relatively heavy. "Black ironstone," the purest form, is a black, fine-grained blocky rock consisting of siderite, iron oxides, sand, clay and coal in varying proportions. "Clay ironstone," a less pure form, tends to be more argillaceous and brownish in color.

Siderite ironstones also occur as a result of complete replacement of calcium in limestones, often with good preservation of original sedimentary textures. A common example is oolitic siderite.

4.10   SILICATES

The most important silicate in sedimentary ironstones is chamosite (hydrated ferrous silicate). This mineral, mixed with other iron minerals (most commonly siderite), sand and mud, forms the most common ore quality ironstones. The rock consists of a primary impure chamosite matrix containing rhombs or aggregates of siderites. Such rocks often contain defined sedimentary character similar to limestones (e.g. oolites), and it is believed that they are true sedimentary in nature and not replacements of primary calcite.

Glauconite, an iron/potassium mica mineral, is a common cement and accessory in sandstones. On occasions it may be sufficiently abundant to require the rock to be described as an ironstone.

4.11   OXIDES AND HYDROXIDES

Ghoethite (basic iron hydroxide, FeO.OH) is often found in oolites, forming concentric interlayers with chamosite. The oolites are present in a fine-grained ground mass of chamosite, clay, siderite or calcite.

Hematite (ferric oxide) and magnetite (ferous-ferric oxide) occur in similar rocks both as oolites and as isolated crystals. It is believed that these minerals are replacements after primary ghoethite.

Limonite (hydrated ghoethite) occurs both as oolites and as larger lumps and nodules in marsh and lacustrine deposits. This "bog iron-ore" is presumaby primary in nature.

4.12 SULFIDES

Pyrite (ferrous sulfide) is common in sandstones and mudstones and may be so abundant as to merit the term ironstone. In black euxinic shales, large aggregates or lenticular bodies of pyrite or marcasite (a polymorph) may occur.

4.13 ALUMINOUS ROCKS

Laterite and bauxite are clays enriched in ferric and aluminum hydroxide. They are a product of sedentary processes in shallow or exposed soils, but may also be transported and detrital in character.

They are dirty white to grey, with brown, yellow or red tinges, massive, earthy and granular, but upper surfaces which have been exposed may be pisolitic or have a hard clinker-like crust. They consist of varying proportions of ferric and aluminum hydroxide, laterite being richer in iron and bauxite in aluminum. Transported laterites tend to have a higher proportion of clay than the sedentary types.

The classic model for production of laterites and bauxites is that of a seasonal tropical climate. During dry seasons, ground waters rich in ions move upward. At surface, evaporation leaves the soil mineral-enriched. However, an alternative model involves the leaching of surface soils by downward-moving surface waters, resulting in enrichment at some depth below surface. Yet another theory suggests that certain deposits may result from the upward movement of hot water of deeper igneous origin.

4.14 PHOSPHATIC ROCKS

Phosphatic sedimentary rocks are relatively common although usually regionally limited in extent. The major source of the phosphorus contained in the rocks is commonly organic. However, various diagenetic and alteration processes often make identification of the original organic component difficult. This is especially true since, unlike limestones, phosphatic rocks are predominately derived from soft parts of organisms which do not leave readily identifiable fossil records.

Most ancient phosphatic sediments are seen in the form of phosphorite. This is a dark bluish to greenish black or brown rock, earthy or compact, that is often nodular or concretionary in structure. The rock consists predominately of apatite (calcium phosphate) with quartz and clay and trace quantities of organic material, dolomite, calcite, iron oxides, and other accessories.

Phosphorite and phosphatic rocks have many diverse modes of origin: primary, secondary, and by the phosphatization of sediments adjacent to primary phosphate deposits. This leads to difficulty in determining environment and type of deposition.

The most common origin of phosphatic rocks is from organic detritus, e.g., coprolites, bone beds, guano. However, such deposits are regionally limited in extent and cannot explain the relatively large phosphorite deposits which are seen. Extensive thin beds of phosphorite often occur, suggesting quiet-water sedimentation. Such sediments possibly result from the anaerobic euxinic envinronment which also produces black sulfurous mudstones. However, such phosphatic deposits are capable of surviving transport and redeposition, thus bringing envinronmental conclusions into doubt.

Another mechanism for the formation of phosphorite is the phosphatization of limestones. Studies show that aragonite is much more susceptible to this than calcite, and evidence of oolites and shells partially or totally replaced by apatite indicates that this process occurs early after sedimentation under conditions where cold phosphate-rich seawater is upwelling into shallower, warmer zones.

Finally, phosphatic nodules are also found in deep water, noncalcitic and apparently transported sediments. No mechanism is presently available to explain this occurrence.

4.15 SALINE ROCKS

This group consists of the rocks which result from the evaporation of brines in enclosed lakes and lagoons. It includes not only those mineral species precipitated from the saturated waters, but also the eventual products contained in an ancient sediment deposited in such an environment.

Enclosed bodies of water formed by the isolation of an arm of the sea evaporate to produce a uniform evaporite sequence. The order of precipitation and relative amounts of each mineral reflects both the composition of seawater (which is regionally and geologically consistent) and the relative solubility of the various components. Figure 4-1 shows the order and amount of precipitation from modern seawater. If minor communication with the ocean continues, sufficient to allow water influx but not recirculation, a stable point will be reached where a single species is being precipitated and renewed. This may lead to great thickness of that salt but will not affect the eventual filling of the basin and the order of precipitation. Minor cyclical variations due to seasonal changes in temperature and water level may be observed.

Where evaporation takes place in an enclosed inland lake or in an isolated arm of the sea where river water continues to enter but where no exit is possible, the resulting evaporites are more variable in composition and sequence. The evaporites formed will be dependent upon the composition and concentration of the lake and entering waters. Seasonal variations in temperature and influx water volume may result in a cyclical sequence of evaporites.

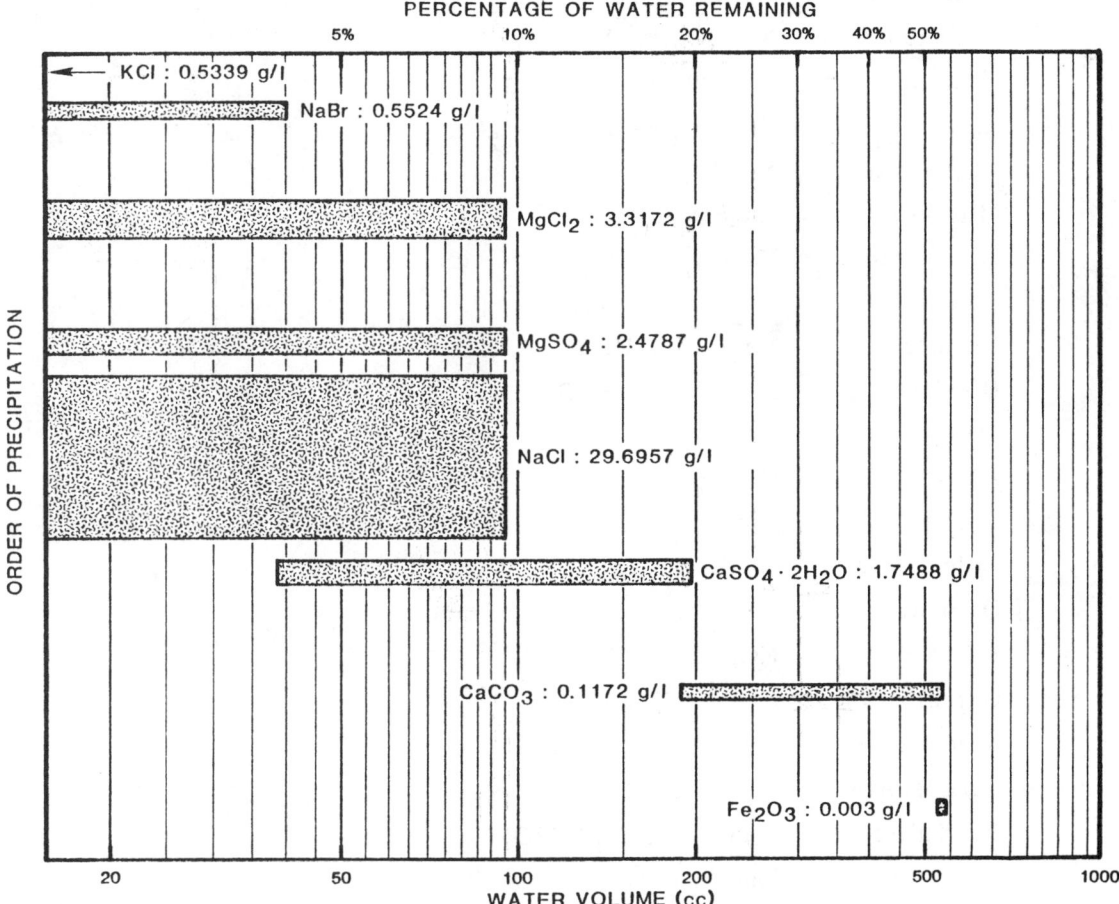

Figure 4-1. Evaporation of Seawater

Although the most volumetrically important evaporite is halite (sodium chloride), many others may occur depending upon the source of minerals in the drainage area. The following list includes the most important:

Mineral	Formula
Halite:	$NaCl$
Thermonatrite:	$Na_2CO_3 \cdot H_2O$
Natron:	$Na_2CO_3 \cdot 10H_2O$
Trona:	$Na_2CO_3 \cdot NaHCO_3 \cdot 2H_2O$
Thenardite:	Na_2SO_4
Mirabilite:	$Na_2SO_4 \cdot 10H_2O$
Glauberite:	$Na_2SO_4 \cdot CaSO_4$
Borax:	$Na_2B_4O_7 \cdot 10H_2O$
Kernite:	$Na_2B_4O_7 \cdot 4H_2O$
Ulexite:	$NaCaB_5O_9 \cdot 8H_2O$
Calcite:	$CaCO_3$
Anhydrite:	$CaSO_4$
Gypsum:	$CaSO_4 \cdot 2H_2O$
Colemanite:	$Ca_2B_6O_{11} \cdot 5H_2O$
Dolomite:	$CaMg(CO_3)_2$
Sylvite:	KCl
Carnallite:	$KCl \cdot MgCl_2 \cdot 6H_2O$
Kainite:	$KCl \cdot MgSO_4 \cdot 3H_2O$
Langbeinite:	$K_2SO_4 \cdot 2MgSO_4$
Polyhalite:	$K_2SO_4 \cdot MgSO_4 \cdot 2H2O$
Kieserite:	$MgSO_4 \cdot H_2O$
Magnesite:	$MgCO_3$
Celestine:	$SrSO_4$
Pyrite:	FeS_2

Evaporite deposits often form readily recognizable sequences of mineral with deposition in inverse order of solubility, the least soluble minerals being deposited first and overlain by layers of increasing solubility. Figure 4-2 shows an ideal sequence of evaporites following this pattern. Initial isolation of the basin from the open marine environment results in a change from normal marine to euxinic sediments, black sulfurous and phosphatic muds. Further isolation and evaporation leads to a precipitation of calcium carbonate, anhydrite and halite in the sequence and quantities seen in Figure 4-1. In the final stage of basin evaporation, the so-called "bittern fluid" will crystallize salts of magnesium and potassium. At this stage complex interactions between the developing crystals and the concentrated brine results in an admixture of minerals and the formation of such complex minerals as polyhalite. Therefore, minerals formed during this late stage may not

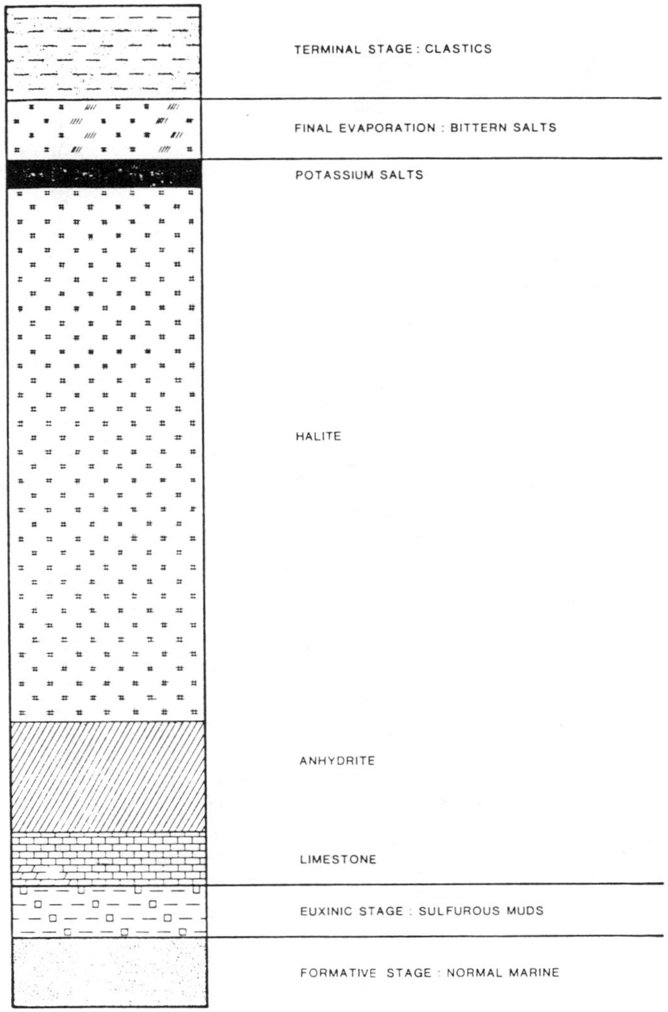

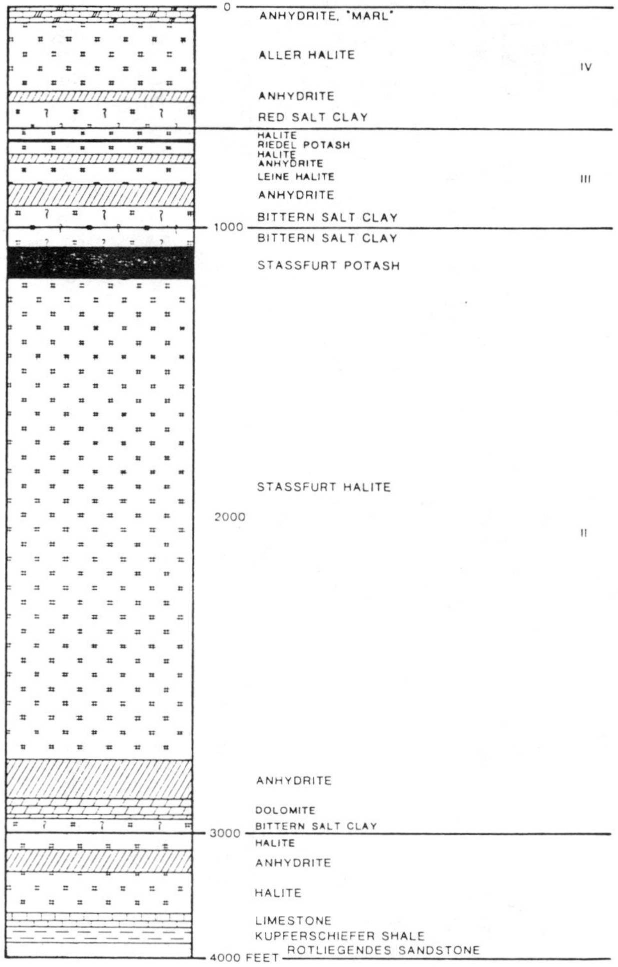

Figure 4-3. Zechstein Evaporite Section

Figure 4-2. "Ideal" Evaporite Section

be deposited entirely according to solubility. The remains of the dispersed sediment load are mixed with and overlay these bittern salts.

In an actual basin, frequent minor changes in ambient conditions (water supply, level and sediment load, temperature, currents, mineral and water densities and solution of evaporites due to fresh water influx) complicate this simple sequence. Zones rarely contain pure minerals but will always be mixed with traces of accessories and detrital material. Boundaries may be transitional or sharp and unconformable. Figure 4-3 gives an example — the Zechstein evaporite section in Northern Europe. In this example, four almost complete cycles of evaporation are seen and, although somewhat modified, the similarity to ideal sequence in Figure 4-2 can be seen.

For purposes of wellsite description and evaluation it is sufficient to separate the evaporite sequence into three subgroups:

- The low solubility calcium salts, gypsum and anhydrite and including the limestones

- The volumetrically predominant halite (structurally significant in the development of petroleum reservoirs)

- The highly soluble and mineralogically complex bittern salts.

When a massive evaporite sequence is to be drilled, it is common practice to set casing and convert to a salt-saturated mud system. If this is not done, the drilling mud will saturate itself by solution which may result in excessive hole enlargement and destabilization of the drilling mud and exposed uphole argillaceous rocks. Alternatively, when highly soluble and reactive bittern salts are to be drilled, then oil-based, emulsion, synthetic, or potassium-based drilling fluids may be used.

In any of these cases good sample recovery will be maintained. The geologist may treat samples in a normal manner. Use a lightly rinsed sample to view the content of evaporate minerals which should retain good crystal habit. A well washed or blended sample should be used to study the insoluble clay or clastic fraction.

A thin or only partially saline formation may be drilled without prior adjustment of the drilling fluid. In this case, most of the soluble minerals will be lost to solution prior to the cuttings reaching surface. The geologist should be careful not to miss important secondary evidence indicative of evaporites:

- An increased and smooth rate of penetration
- Decreased volume of cuttings
- Eroded or reworked appearance of cuttings
- Increased drilling fluid salinity and conductivity
- Increased drilling fluid flowline temperature
- A generally declined and smoothed background ditch gas, possibly with occasional sharp, well-defined shows
- Salty encrustations on the surface of dried samples

4.16 CALCIUM SALTS

4.17 Carbonates

Initial evaporation of the basin with some increase in salinity will not produce immediate crystallization. Figure 4-1 indicates that almost 50 percent of the water volume must be lost before precipitation can occur. In the warm, quiet water enriched in ions, normal marine limestone formation occurs — producing fossiliferous micritic limestones.

At the basin margins in the tidal and supratidal zones, more rapid evaporation may result in the precipitation of gypsum and in the formation of a saliferous sabkha. The removal of calcium from supratidal ground waters results in relative magnesium enrichment and penecontemporaneous dolomitization of the adjacent marine limestones by the process of evaporative reflux (see paragraph 3.53).

It is unlikely that a natural basin will exhibit a smooth bottom profile, deepest at the center and shallowing uniformly toward the margins. Most basins consist of a major asymmetric basin with variable water depths and perhaps satellite basins or embayments separated by seabed highs. Often, communication with a larger marine body or ocean is via a narrow channel of similar water depth to the basin, but more commonly a broad shallow platform will exist — providing a forereef or forebasin feature, limiting the flow of water into the basin and perhaps halting it totally during periods of seasonal low water level.

Like the basin margins, points of structural high will have a higher-than-normal evaporation rate since the shallow water depth results in a high surface-to-volume ratio, increasing both the rate of heating and evaporation of the water. Higher rates of evaporation encourage deposition of calcite and dolomite at these highs, causing accentuation of the features and further separation of the basin by "reefs" or swells. Denser concentrated brines formed at these points settle to deeper portions of the basin where they may accumulate or pass from the basin via permeability to be replaced by new seawater influx. In the depths of the basin, lack of circulation and oxygenation provides euxinic conditions. Laminated black micritic dolomites and bituminous, pyritic shales and silts are predominantly barren of fossils.

At this stage in the basin's development, three distinct environments are recognizable: (1) the sabkha, (2) the relatively normal marine, and (3) the euxinic. With seasonal changes in water level and the eventual shrinkage of the basin, each of these may be transgressive or regressive of the others. Further complications are added during diagenesis by complex interactions between the minerals (e.g. gypsum/anhydrite conversion, dolomitization, de-dolomitization, calcification of clays, etc.). The eventual rock is commonly a recrystallized micritic limestone, variously dolomitized and speckled with anhydrite. Mutual pseudomorphing of calcite and dolomite commonly occurs.

4.18 Sulfates

Gypsum and/or anhydrite commonly are formed in the supratidal zones, and structural highs from evaporation and concentration of ground waters. However, evaporation at the surface of a static water body may produce a concentrated surface layer sufficient to precipitate calcium sulfate in deeper parts of the basin. Like the calcitic dolomites, the anhydrites tend to be thickest where the basin is shallow and will further reinforce these swells. With increased evaporation, a more general deposition of sulfates occurs throughout the basin. There is much discussion as to whether the original mineral deposited is gypsum or anhydrite. This results from the ready and rapid inversion of the two minerals, which will be discussed below. A more immediate problem is that many wellsite geologists fail to discriminate which of the two is <u>currently</u> present, or worse, arbitrarily decide upon a mineral name on the basis of depth or age of the sample (e.g., gypsum in young rocks, anhydrite in older rocks.) Although discrimination between anhydrite and gypsum is not always possible at the wellsite, an attempt should always be made. Figure 4-4 tabulates the differences and similarities of the two minerals.

CHARAC-TERISTIC	GYPSUM $CaSO_4 \cdot 2H_2O$	ANHYDRITE $CaSO_4$
Color	White, light to dark gray, red, blue, yellow, brown	White, pale gray, red
Structure	(1) Selenite crystals, glassy, slightly flexible, fibrous structure (2) Satin spar, fibrous-lacy, pearly opalescence (3) Massive, fine-grained, subvitreous to dull luster (4) Spongy, white, soft	(1) Fibrous, parallel and radiate structure, fine-grained (2) Amorphous, fine-grained, massive but cleavable
Luster	Pearly, subvitreous, earthy	Pearly, greasy, vitreous
Hardness	1.5 to 2.0 on Moh's scale, can be scratched by fingernail	3.0 to 3.5 on Moh's scale, can be scratched by brass pin
Density	2.30 to 2.37 g/cc	2.9 to 3.0 g/cc

Figure 4-4. Characteristics of Gypsum and Anhydrite

Even if discrimination is not entirely possible, the presence of a sulfate mineral can be confirmed by testing with barium chloride solution.

1. Place three cuttings in a test tube or cut bottle and fill with distilled water.

2. Agitate and pour off water. Refill and repeat.

3. Half fill with distilled water and add three drops of dilute hydrochloric acid. Agitate.

4. Add two drops of barium chloride.

5. A pearly white discoloration will confirm the presence of gypsum or anhydrite.

It is now generally agreed that gypsum alone is present in the environment of deposition. Chemical studies and observation of modern environments (e.g. the Persian Gulf) indicate that, even in the extreme conditions of temperature and salinity (which would favor anhydrite formation), metastable gypsum is more commonly formed.

Gypsum formation in supratidal sediments is most commonly seen in modern environments. If the sediment is soft and unconsolidated, euhedral crystals will grow by displacing the host sediment. Displacement cannot occur in a rigid host rock, e.g. detrital sediment or a cemented carbonate originating in the basin. Gypsum crystal growth begins in pore spaces and continues in random orientation, often containing included particles of the host sediment.

Although rare in modern sediments, there is strong ancient evidence for the deposition of gypsum from concentrated brines within a standing body of water. In order for such a process to produce the great thicknesses of layered sediments, a delicate balance must have been maintained between evaporation and precipitation rate and the influx of fresh water sufficient to supply fresh ions, but not so great as to dilute the basin and prevent further precipitation.

That such conditions can occur is evidenced by laterally extensive laminated deposits showing bedding and other sedimentary structures. The seasonal or episodic nature of sedimentation is marked by the interbedding of anhydrite or gypsum layers with detrital clays, limestones or dolomites. Such cyclical sediments, if truly seasonal in origin, are referred to as "varves".

Very soon after burial, conversion to anhydrite results. The actual depth of burial required for conversion to occur is difficult to determine, being a function of overburden, geothermal gradient and pore water volume and concentration, but it would appear to be in the region of 1000 feet or less. When this depth is reached, the complete change from gypsum to anhydrite occurs over a few tens of feet.

The significance of the change from deposited gypsum to lithified anhydrite is that, in addition to physical compaction, there is a 38 percent reduction in solid volume in the conversion process, but an overall increase in bulk volume; i.e., the volume of water plus anhydrite is greater than the original volume of gypsum. The result will be abnormally high fluid pressures which, if accompanied by low permeability, cause the sediment to "flow" and thus create deformation structures and dislocations. If outflow of this excess fluid is possible, the calcium-sulfate-saturated brine will be displaced (both laterally and vertically) considerable distances and precipitate anhydrite in sediments which originally had no part in the evaporative environment.

Anhydrite generally grows in a microcrystalline form. Where sufficient dissolved calcium sulfate is available to replace the 38 percent loss in volume, microcrystalline aggregates may accurately pseudomorph after original gypsum crystals. Usually gypsum crystals will be replaced by irregular nodules of anhydrite. Bedded gypsum will be replaced with similarly bedded anhydrite with little loss of structure though with inevitable shrinkage. Figure 4-5 shows this process.

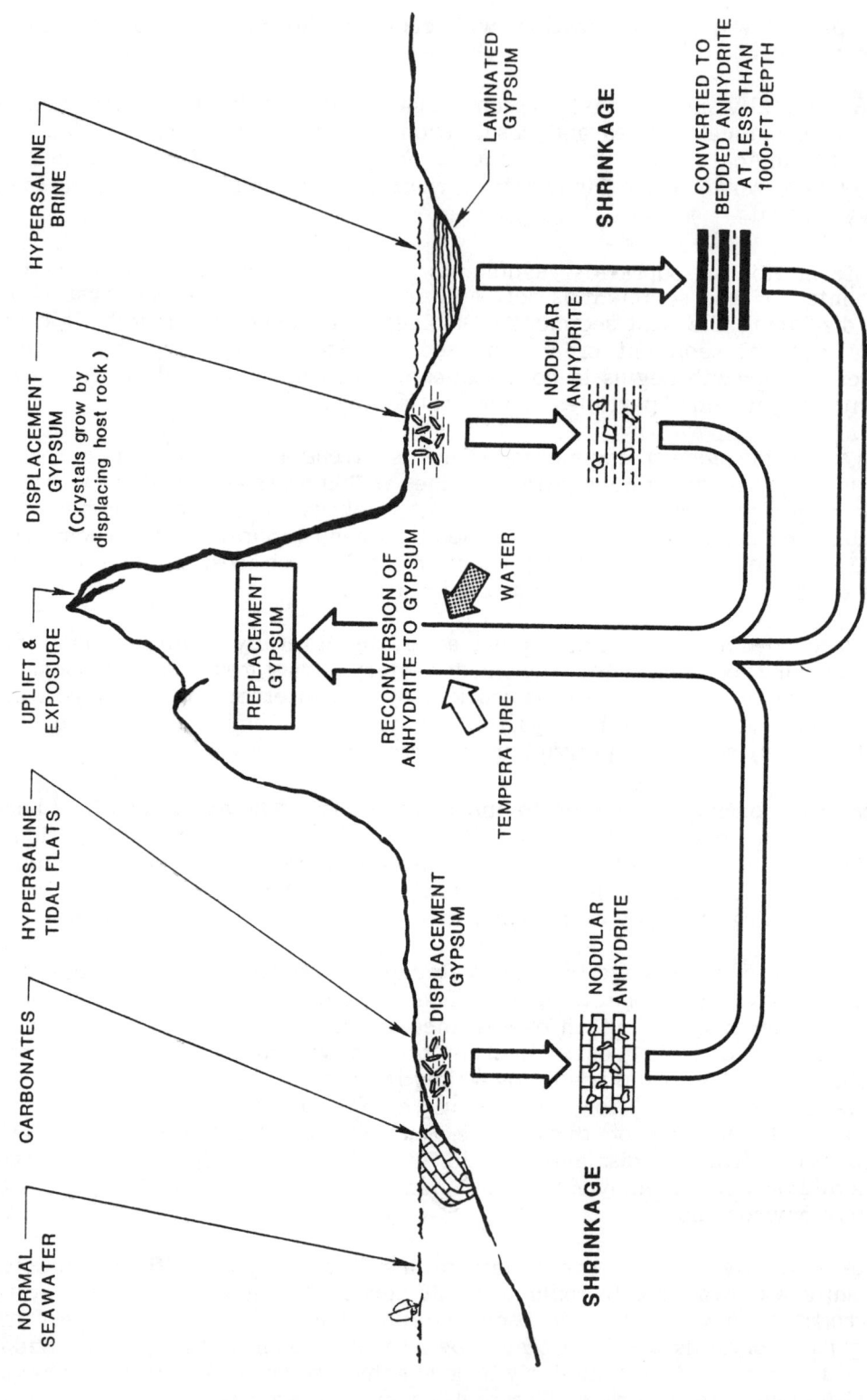

Figure 4-5. Types of Gypsum-Anhydrite Conversion

If the sediment is retained at the depth of conversion, or if later uplift and erosion return it to these conditions, reconversion to gypsum may result. The process is dependent upon the availability of plentiful supplies of water, the salinity and temperature of that water. Since these are highly variable, the reconversion depth is even less definable than the original conversion depth such that secondary gypsum may occur as deep as 9000 feet while some anhydrites may be exposed unchanged at surface.

4.19 HALITE

The formation of halite requires similar though more extreme conditions than those which result in the formation of gypsum. Sufficient evaporation is necessary to reduce the water volume in the basin to one-tenth the original to allow precipitaton of sodium chloride. If thick sequences are to be produced, an influx of new seawater is required as well as some outflow of the heavy bittern salt brine. This may be possible by some narrow restricted channel or by a reflux process through permeability in the bottom or flanks of the basin. In either case, evaporation and crystal formation occur in the heated, refreshed upper layers of the water. Denser concentrated brine sinks to the bottom of the basin to be drained via a channel or seafloor permeability.

Halite formation initially occurs in the deeper parts of the basin where heavier brines are accumulating and there is no drainage or loss through permeability. In such circumstances, halite precipitation from heavy brines may be occurring at depth while gypsum and dolomite precipitation continue to occur in the shallow refreshed surface brines and at the basin swells. In this way and with periodic environmental changes, some transitional, cyclical or patchy sediments may result.

Halite sections may be several thousand feet thick. Halite occurs as massive crystalline beds showing good cubic cleavage. It is colorless to white, often with a pink or red tinge, and is readily recognizable by its cubic habit, solubility and taste. It is possible to test for halite using silver nitrate solution which responds to the presence of the chloride ion by forming a white precipitate. This test is not conclusive since most formation waters and drilling fluids are saline and will also form a precipitate with nitrate. Rigorous washing with distilled water will remove mud filtrate but may also dissolve any contained halite. A more conclusive test is to monitor increasing mud salinity using a quantitative silver nitrate titration. (See Appendix B of the Field Geologists Training Guide, (EXLOG, 1985).)

Halites are rarely pure, commonly being associated with interbeds, bands or dispersed red clays. Thin beds and lenses of anhydrite or gypsum are also common, indicating episodic environmental changes. Small amounts of sodium bromide will replace sodium chloride throughout the salt body, the initial ratio of sedimentation being a function of relative concentrations and solubilities of the bromide and chloride ions in the concentrated brine. In theory, the bromide concentration in a halite section should increase uniformly from bottom to top, that is, with increasing concentration of bromide in the brine. In fact, environmental changes during deposition and resolution and redistribution after sedimentation tend to destroy this uniformity.

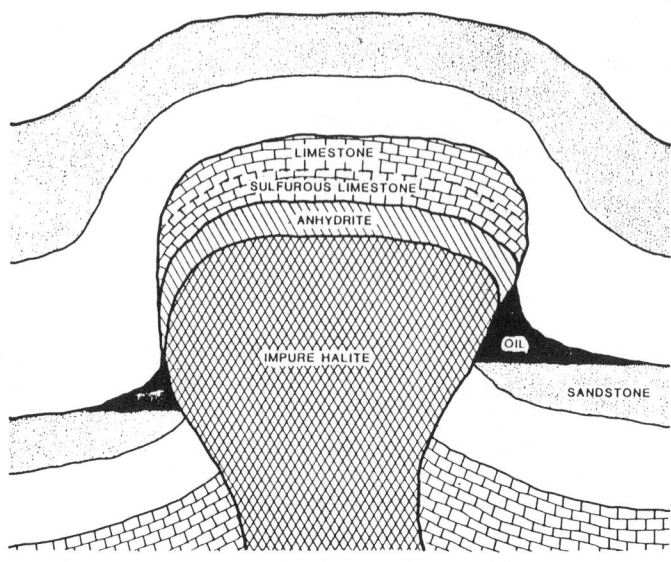

Figure 4-6. Typical Salt Dome

The most significant characteristic of halite, in terms of petroleum geology, is its tendency to flow under pressure rather than fracture as do other rocks. Deformation of salt beds may result in vertical thickening many times that of the original bed. Where shale or anhydrite beds exist, the increased separation of these may be used as a measure of bed deformation.

Distortion of overlying beds by salt deformation can result in the formation of anticlines or domes in which petroleum accumulation may take place. Further deformation may result in the salt dome piercing overlying beds to form a diapir in which the salt both defines the structure and provides a seal or cap rock for the petroleum accumulation (see Figure 4-6). The salt will normally be observed to have a "cap rock." This consists of low-solubility residues, predominantly anhydrite, resulting from the solution of the upper part of the salt diapir by circulating ground waters. Petroleum desulfuring processes and the action of sulfate-reducing bacteria often result in the reduction of the anhydrite to calcium carbonate and elemental sulfur.

4.20 BITTERN SALTS

In commercial salt production, after the major evaporation and precipitation of halite is completed, a dense (1.31 g/cc) brine remains, containing the most soluble components. This "bittern liquor" consists mainly of magnesium salts and chlorides and bromides of potassium and sodium, and is normally flushed away.

In a natural evaporative basin (where complete isolation exists) the bittern liquor evaporates to produce a group of potassium/magnesium bittern salts. Due to the commercial significance of potassium and the mining of these salts as a source of it, they are very commonly given the general title "potash salts." While this term is so widely used as to be acceptable in wellsite descriptions, the geologist should remember that what is simply described as potash salt is in fact a very complex mixture of potassium and magnesium double salts.

In order for a thick halite sequence to form, an interchange of water between the evaporite basin and an open marine (or at least saline) environment is required. By reflux or communication, dense bittern brine must be flushed out and replaced with new brine (see paragraph 4.19). In such an environment, potash salts would never be produced.

The presence of a shallow forebasin or broad shallow channel between the main evaporative basin and the open sea will allow seawater to enter but will serve as a barrier to heavy brine outflow. Such a barrier may be structurally present in the basin at origin or may develop due to the marginal buildup of dolomites and gypsum.

Many basins which fulfill the above condition do not contain potash salts. This is because of the high solubility of the salts. Even minor water flow through the basin before or during burial will result in their complete dissolution and loss from the basin. Alternatively, many basins are seen in which several complete sequences of evaporites occur, each topped by potash salts (Figure 4-3). This indicates that complete evaporation of the basin has alternated with rapid influx of new seawater and a new cycle of evaporation and precipitation quickly enough to prevent the resolution of the potash salts.

The lateral distribution of potash salts within a basin is complex. This is a function of physical characteristics of the basin and chemical changes in the bittern brine itself and in the whole evaporite section after deposition. These physical and chemical factors are strongly interrelated before, during and after deposition and must be considered together.

Where several interconnected basins occur, halite, potash salt or both may be deposited in the depths of main or satellite basins while brine concentration and gypsum deposition are taking place at the forebasin and at the shallow swells. Alternatively, periods of reduced water influx may result in exposure of the swells, isolation of the basin segments and cycles of evaporation and deposition. In such conditions, refilling of the basin would be slow and, having entered the deep basins via the shallow, the inflowing water would already be somewhat concentrated — preventing major resolution of the previously deposited salts. In such a complex basin, marginal deposits of less concentrated brines would be displaced laterally (toward the central deeper basin) by precipitates from increasingly more concentrated brines which, with final closure of the basin, would also succeed them vertically. Subsequent refilling of the basin would repeat the cycle both laterally and vertically. This simple picture may in fact be complicated by the settling of shallow precipitates to depth through density layering of brines of differing concentrations, or may be the development of locally hot or cold, dense or light zones due to convection currents and basin structure.

Within the deep, concentrated parts of the basin, the onset of potash salt precipitation is marked by the presence of polyhalite (calcium.magnesium.potassium sulfate) precipitating with halite. This occurs when the brine has been concentrated twenty times and continues until the brine becomes saturated — when true potash salts displace both halite and polyhalite.

Polyhalite rarely occurs in large enough aggregates to be individually recognizable. It must be determined by the changed appearance it gives to the halite with which it is intimately associated. It is strong brick red in color and is therefore similar to the reddish clays which are also commonly associated with halite. The clays, however, tend to be banded or unevenly distributed through the salt, and tend to produce a translucent "dusty" luster. Polyhalite will give an even red coloration to the halite without affecting its transparency or luster. Most notable is the bitter "peppery" taste which it gives to the salt, and, if sufficient section has been drilled, a characteristic unpleasant sour/bitter odor. This will be noticed in the mud and around the shale shaker. (The odor has been likened to that of a stale locker room! This description, though unpleasant, is most appropriate.)

Kainite (magnesium sulfate.potassium chloride) and carnallite (magnesium.potassium chloride) are the first potash salts to form — kainite in the shallow warmer surface water, and carnallite in the cooler depths. Sylvite (potassium chloride) may be formed at surface by intense evaporation, but, in settling through the deeper magnesium-rich brines, it (and part of the kainite) is converted to carnallite. At the time of deposition the majority of the potash salt deposit consists of carnallite with some kainite.

During compaction and diagenesis, many complex changes result — both chemical changes and physical displacement of the potash salt. The major cause of the potash salt diagenesis is the large quantities (4.86 cu m per 10 cu m) of sulfate-rich water released during the conversion and compaction of gypsum to anhydrite. Upward and laterally moving water migrates through overlying halite through stress fractures and by solution and in the process, becomes saturated in sodium chloride. Where carnallite overlies the halite it will be totally dissolved and replaced by halite. Carnallite deposited in structurally low positions in the basin may be bypassed by the migrating groundwaters and remain unaffected.

The dissolved carnallite will be carried updip. Some separation may occur, especially where the carnallite is sulfate-rich, resulting in precipitation of kieserite (magnesium sulfate) downdip and sylvite (potassium chloride) at the structural highs. In general, however, a complex mixture of the potash salts and some halite results.

The potash salts are similar to polyhalite in color and odor. They rarely occur in a well-developed crystalline form, usually being seen in massive microcrystalline aggregates. Due to the updip migration following deposition, they are commonly found indurating overlying sediments or infilling the intergranular porosity of the dolomite and anhydrite of subsequent evaporite series.

4.21 CARBONACEOUS ROCKS

Apart from the bitumens or solid hydrocarbons discussed elsewhere, (Mud Logging: Principles and Interpretations (EXLOG, 1985) and paragraph 4.27), the most important sedimentary carbon-bearing rocks are the coals. These are bedded rocks formed by the accumulation of vegetable material. In wellsite geology, coal beds provide useful marker horizons visually in drill rate, in gas analysis (they give small, well defined methane shows, allowing a useful check of gas lag versus cuttings lag), and on the gamma ray and formation density logs. They also may serve as source rocks of nonassociated gas reservoirs. Even small quantities of carbonaceous material in other sediments may be of value in geochemistry in assisting determination of thermal history.

Coals occur in beds of extremely consistent thickness and often of major regional extents. Although tens of feet thick at deposition, decomposition and dewatering during burial result in a typical thickness of six feet or less.

Coal beds often occur in sequences interbedded with detrital sediments, indicating a cyclical downwarp and infilling of a low-lying marsh basin followed by rapid subsidence and infilling with sediment. When sediment fill returns the basin to its original elevation, revegetation and a new cycle of plant sediment begin. This mechanism is supported by the fact that coal seams are commonly very clean, consisting of almost pure plant material with little detrital sediment. Detrital rocks below coal beds, so-called "seat earth," show extensive plant root remains and perturbation. All of this strongly supports the theory that coals are authigenic deposits, accumulated in place with little or no transport.

Coals resulting from transported plant material do occur and are called cannels. These are dull, lusterless grey coals forming lens and channels in normal coals. They are formed largely of spores and often contain fish scales and other fossil material.

4.22 CLASSIFICATION

Following burial, increases in temperature and pressure have a major effect, inducing major chemical changes in the vegetable matter. The changes essentially involve reduction of the organic material and may be likened to combustion in an oxygen-depleted environment. Biochemical decay stimulated (and later replaced) by high temperature results in the complete breakdown of the organic compounds to elemental carbon and volatile components. The volatiles and water are driven off, leaving a rock consisting predominantly of amorphous carbon. The process, which is sometimes referred to as "intramolecular combustion," is illustrated in the following equation:

$$C_6H_{10}O_5 \longrightarrow CO_2 + 3H_2O + CH_4 + 4C$$

cellulose   carbon dioxide   water   methane   carbon

The process is not this simple, cellulose being the most important but only one of many compounds contained in plant matter. There are several intermediate breakdown processes, and the conversion is a gradual and continuous one occurring through all stages of coal diagenesis. Some volatiles, e.g. methane and carbon dioxide, may remain trapped in the rock by loss of permeability with compaction. In mines, accumulations of these gases are referred to as fire damp (methane) and choke damp (carbon dioxide). Sulfur and sulfur compounds may also be retained or may form secondary sulfide minerals such as pyrite.

The passage of coal through the various stages of coalification may be quantified in terms of increasing "rank." For general descriptive purposes, the terms peat, lignite, bituminous coal and anthracite are acceptable. The relationship of these terms to rank and the chemical composition of the coal are shown in Figure 4-7.

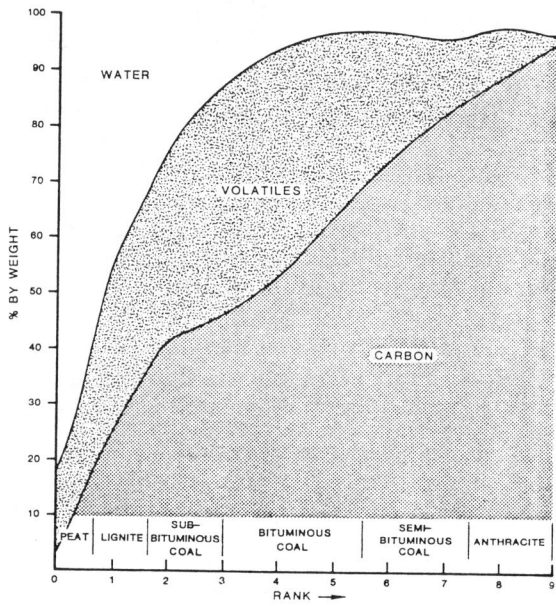

Figure 4-7. Types of Coal

4.23 Peat

The origin of most coals is an accumulation often tens of feet thick of plant debris commonly in a marsh or bog environment. Under superficial compaction and stimulated by the heat generated from biochemical decomposition, the lower layers become partially decomposed and carbon enriched. Peat is seen only as a superficial deposit and is never preserved in ancient rocks other than as a coal of a higher rank.

4.24 Lignite

Lignite or "brown coal" is found in geologically young deposits. It is firm and well compacted but shows well-developed vegetal structures and contains up to 40 percent water. It is easily recognized in cuttings but it may be impossible to isolate lignite from the formation from lignite and lignosulfonate mud additives.

4.25 Bituminous Coal

The first true coals are formed by further diagenesis. These are the major and most commonly seen forms of coal; for convenience, they are divided into three subgroups:

- sub-bituminous
- bituminous
- semibituminous

Sub-bituminous coals show the black color and the dense, compact nonvegetal structure of true coals. However, due to their high water content (up to 25 percent), upon drying they have a tendency to crack and crumble — a process known as "slacking."

Bituminous coals ("fat coal" in mining terminology) are compact, brittle, high-luster black coals. They exhibit a lamellar or conchoidal fracture but also show a characteristic perpendicular jointing, giving the common cuboidal shape to cuttings. No vegetal structure remains although occasional plant fossils may be present at shaly partings or bedding planes in the coal.

The strong jointing pattern in bituminous coal is a result of banding in the coal, resulting from original bedding and from segregation during diagenesis. Perpendicular to the parallel bedding planes of the coal are two sets of vertically perpendicular joints (known as "cleats" and "ends" in mining) formed by planes of weakness between bands of different coal types.

The major matrix of the coal is made of the material vitrinite. This is dense, glass- or gel-like microlithic carbon. It is hard, brittle, and has a high luster. Although plant cellular structure may be visible in thin section, in normal observation it is effectively structureless. The availability of vitrinite to reflect incident light, a means of quantifying luster, is closely related to its order of diagenesis and hence the maximum temperature to which it has been subjected. Thus in geochemistry, vitrinite reflectance of coals and of carbonaceous particles in other sediments may be used to reconstruct paleogeothermal gradients and identify the maturity of petroleum source rocks. This procedure is normally performed in a laboratory, but on occasion may be done at the wellsite.

Bituminous coal consists of bands and layers of vitrain which is almost pure vitrinite and clarain which contains resistant plant debris in a matrix of vitrinite. The rank and commercial quality of the coal will increase with the percentage of vitrain, the high being semibituminous coal (steam coal) which has less clarain and commonly less volatiles and water.

Along the bedding planes in coal there tends to accumulate a granular residue of insoluble plant residues and detrital material. This material, known as fusain, is porous — allowing the outflow of excluded water and volatiles. This is the location of gas and secondary minerals (such as pyrite and calcite) in bituminous coal.

4.26 Anthracite

The highest rank of coal is found only where temperatures and/or pressures occur in excess of those present in diagenesis. Where coal beds have been extensively deformed at depth or adjacent to plutonic intrusions, anthracite may form. Anthracite consists of almost pure carbon with little water, volatiles or residual material. It is hard, brittle, and has a high vitreous or submetallic luster due to its high vitrinite content.

The decorative mineral, jet, is an anthracitic form of cannel coal. Although similar in appearance to anthracite, it is less brittle and may be worked or rounded.

4.27 SOLID HYDROCARBONS

Solid or semi-solid hydrocarbons may occur in reservoirs, potential reservoir rocks, fractures, and petroleum migration pathways. They are chemically similar to petroleum, especially to the solid residuum remaining after distillation. Like petroleum, and unlike coals, they show evidence of migration from their point of origin.

Solid hydrocarbons may result from the migration and solidification of immature petroleum or from the fractionation of mature petroleum during migration or at seepages. Some types suggest metamorphism and dehydrogenation or polymerization after solidification.

As complex mixtures, solid hydrocarbons are infinitely variable, and many distinct types have been named. At the wellsite it is normally sufficient to recognize the three basic types:

- Mineral waxes
- Asphaltites
- Pyrobitumens

4.28 Mineral Wax

Waxes are commonly relatively pure paraffin hydrocarbons. They are soft, amorphous, range in color from golden yellow to dark brown and are often semi-transparent.

Waxes often have a low intensity, light colored fluorescence under ultraviolet light. They are soluble in organic solvents, resulting in a similar or brighter cut fluorescence. A wax residuum in a water-bearing zone may therefore be misinterpreted as a true oil show.

Careful cuttings examination may reveal the true situation. Other factors which will indicate the falseness of the show will be

- Lower-than-expected gas show
- Patchiness of fluorescence
- Dullness of fluorescence
- Slower or streaming cut

4.29 Asphaltites

Asphaltites are similar to natural asphalts or tars but have a higher melting point than their surface equivalents. They are black or dark brown in color with a bright luster. They are brittle and form a conchoidal fracture, but may be plastic or greasy when probed and may soften with gentle heating.

Alsphaltites often occur as discrete nodules or "blebs" in pores or fractures. Their fluorescence may range from dull brown spotted to none under ultraviolet light, but they may show a weak to moderate intensity, slow, streaming white to yellow-white cut fluorescence. It is extremely important to locate and identify the asphaltite.

4.30 Pyrobitumens

Pyrobitumens are similar to asphaltites, though less common, and in trace quantities may be indistinguishable. They have a lower luster and will not soften, but will decompose and burn on heating.

Pyrobitumens have little or no fluorescence and are only poorly soluble in organic solvents.

5
IGNEOUS AND METAMORPHIC ROCKS

5.1 IGNEOUS ROCKS

Igneous rocks are not of major significance in petroleum geology, but when they do occur it is essential that the logging geologist is able to produce a valid description and, if possible, a name. Drill cuttings are not the ideal medium to work with when describing igneous rocks, but certain simple observations such as crystal size and form, proportion of light to dark minerals, and the presence of free quartz all help to identify the rock. If the sample is good enough to allow identification of the type of feldspar present, a complete characterization can often be made.

Igneous rocks rarely form petroleum reservoirs, although such reservoirs do exist in fractured or fragmental igneous rocks. They may constitute important marker horizons; their identification is important in a decision that basement is reached, and they provide a major source of detrital sediments, identification of the origin of which may be critical.

5.2 CLASSIFICATION

Classification of igneous rocks is at the simplest level on the basis of method and location of the original occurrence and emplacement. Beyond that level no systematic scheme of naming is used, and a combination of mineralogy and texture is the criterion for naming a rock. Further discrimination is possible, but without the use of a petrological microscope and geochemical equipment this is not done at the wellsite.

5.3 Plutonic Series

Major intrusive bodies emplaced at depth may result from the intrusion of large bodies of magma or from the melting and recrystallization of preexistent rocks. In either case they are characterized by a uniform crystallinity, generally coarse texture and often banding due to gravity settling or convection. These features are characteristic of crystallization within a large compact body. Originally, it was thought that great depth was a factor in producing such texture, hence the name plutonic (see Hypabyssal Rocks, paragraph 5.8). It has been shown that it is the shape and size of the body which controls the eventual rock type and texture, so the name, while still used, is something of a misnomer.

Since by nature they are constituents of large igneous bodies, the encountering of plutonic rocks in an exploration borehole is usually indicative that economic basement has been reached. Such bodies commonly cause regional or thermal metamorphism, with resultant loss of permeability and destruction of hydrocarbons in adjacent sediments, further reducing the economic potential of the section. They may of course have major economic significance in exploration for geothermal (rather than hydrocarbon) energy sources.

5.4 Volcanic Association

Igneous rocks emplaced in smaller quantities at or near surface will characteristically be finer crystalline or even vitreous. They may be fragmentary, or contain or be mixed with fragments of preexistent rocks. Although chemically similar in type, they can, on the basis of texture and occurrence, be subdivided into three major types (paragraphs 5.5, 5.6 and 5.7).

5.5 Intrusive: Minor emplacement of igneous rock often follows existing planes of weakness, e.g. bedding planes, joints, fractures. They are rarely of major extent and hence cool rapidly, yielding fine crystalline rocks and little alteration in adjacent sediments.

5.6 Extrusive: Magmatic intrusions may extend to surface and be extruded as lava. The rocks will be essentially similar to small intrusions but may cool so rapidly as to be microcrystalline or to form glass. Layering may be distinguishable, showing several generations of lava flows.

5.7 Explosive: Viscous magma may approach surface, containing large quantities of retained volatile gaseous components (c.f. gas-cut mud). The exsolution of these volatiles at low surface pressure produces a rapid, massive increase in volume. In a closed magma chamber, the result will be a cataclysmic explosion. The debris of this explosion may be regionally very extensive. It will consist of cinders, ash and glass droplets and shards or pumice from the magma "foam." Closer to the source of the explosion, larger fragments of previous rocks (both original sediments and solidified lava) may be found.

Volcanic rocks, although sometimes laterally extensive, do not commonly constitute large volumes of rock. Thus they may be encountered in a sedimentary section without affecting its economic potential either by their own presence or by alteration of the sediments. The most common volcanic rocks encountered in petroleum exploration consist of explosive debris deposited, and sometimes transported, in adjacent marine basins.

5.8 Hypabyssal Rocks

This term, now obsolete, was used to describe rocks intruded at shallow depths. The term was based upon the assumption that the textural differences between plutonic and intrusive volcanics were due to the depth of emplacement. Since it is now known that this is not true, that size and compactness of the igneous body are the controlling influence and that small shallow intrusive rocks are essentially volcanic in nature, the term is superfluous and should not be used.

At the wellsite, working with cuttings or, at best, cores, distinction between extrusive and intrusive (hypabyssal) volcanic rock may not be possible. While texture is important, location of occurrence is also a factor and this may not be immediately apparent.

5.9 DESCRIPTION

Like all rock descriptions, effort should be placed upon making the rock recognizable to some future reader. Since igneous rocks rarely have petroleum significance, little interpretation is necessary beyond finding a name. Only a simple, objective description of obvious physical features is required. The descriptive procedure outlined in this section is not rigorous and would not satisfy the needs of an igneous petrologist. It is intended to assist the petroleum geologist to obtain a rock identification with the commonly available wellsite tools.

5.10 Silica Percentage

Igneous rocks are usually subdivided on the basis of being acid, basic, or further divisions on the same scale. This subdivision bears no relationship, either direct or indirect, to the usage of these terms in chemistry. (An "acid" rock is not acidic, i.e., it does not contain free hydrogen or hydronium ions capable of reacting with a base, or alkali, to form a salt. The term "alkaline" is also misused in igneous petrology; see paragraph 5.23, Feldspars.)

The terms acid and basic, used in reference to an igneous rock, are defined by the amount of the chemical component silicon dioxide reported from a "norm" or "normal analysis" of the rock. Based upon the percentage of silica, the following division may be made:

- acid: more than 66% silica
- intermediate: 52 to 66% silica
- basic: 45 to 52% silica
- ultrabasic: less than 45% silica

It is important to remember that this normal percentage bears little or no relationship to the actual "modal" percentage of quartz present in the rock. A normal analysis requires a complete chemical analysis of the rock for oxygen, the seven other major elements (silicon, aluminum, ferric and ferrous iron, calcium, sodium, potassium, and magnesium), and the trace elements (titanium, phosphorous, hydrogen, etc.). The results of the analysis are normalized by computing the percentage of the oxides of each of the major and trace elements. Such oxides are rarely, if ever, all present in the rock, but this format of analysis results allows the identification of different rocks resulting from chemically similar sources.

Obviously, a normal analysis cannot be performed routinely at the wellsite. A modal analysis (percentage of actual minerals present) by volume may be estimated, and fortunately bears sufficient relationship to the norm to allow a distinction to be made.

Rocks rich in total silica (norm) are similarly abundant in high-silica-content minerals (mode) and notably contain free quartz. Conversely, rocks poor in total silica are richer in the silica-poor ferromagnesian minerals. It is fortunate that discrimination between silica-rich and silica-poor minerals is readily made on the

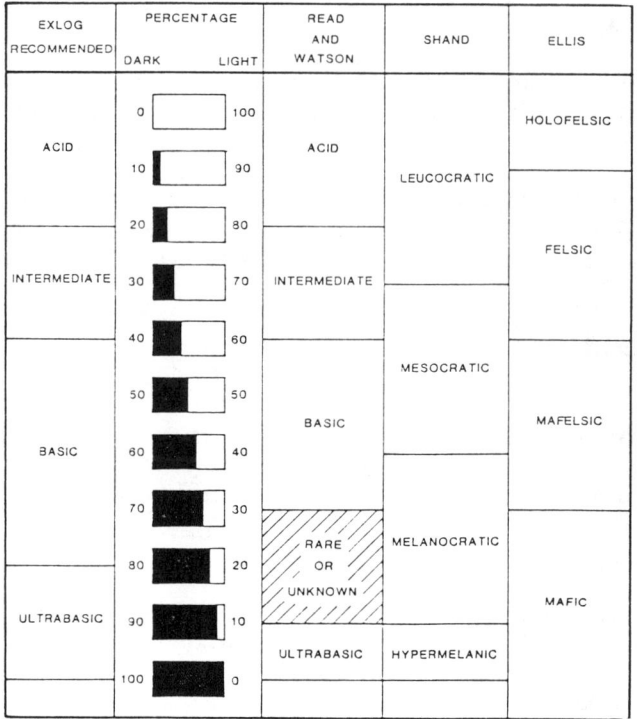

Figure 5-1. Classification by Color Separation

basis of hue, the former being light in color, the latter dark. An estimate may therefore be made on the basis of the proportion of light and dark minerals present in the rock, even prior to microscopic examination (Figure 5-1):

- acid: more than 80% light minerals (including free quartz)
- intermediate: 60 to 80% light minerals
- basic: 20 to 60% light minerals
- ultrabasic: less than 20% light minerals

Thus an igneous rock may at least be classified into one of these subdivisions at the wellsite. Unless the sample quality is very poor, the geologist should not stop at this observation but use it as a guide in proceeding with its description and identification. Thus if a rock is finally identified as a basalt, the inclusion of the term "basic" in the description is superfluous. However, if a rock type cannot be conclusively selected, then the rock must be named a "basic igneous rock" and a full list of observed characteristics given.

The second observation, after color separation, should be the quantity of free quartz present in the rock. The major light-colored minerals present in igneous rock are quartz and feldspars. Quartz is readily distinguished by its greater hardness, lack of color and greater transparency due to the absence of twinning in the fresh mineral and of weathering products in exposed samples. If crystal form is sufficiently developed in the rock, lack of defined cleavage will also distinguish quartz.

Only acid rocks will contain appreciable quantities of free quartz, commonly in excess of 10 percent. Intermediate or basic rocks may contain accessory quantities of free quartz. Ultrabasic rocks will contain less than 20 percent (commonly less than 10 percent) light-colored minerals and no free quartz.

5.11 Grain Size

Most igneous rocks are crystalline in texture, and the size of the crystals is both genetically significant (indicating the cooling history) and an important criterion in selecting the rock-type or name. Like sediments, the modal grain size is a significant key to the rock's history. The presence of crystals of markedly different size to the ground mass may also yield important historical information.

The traditional grain-size nomenclature used for igneous rocks is unfortunately different from the Wentworth scale (Figures 1-4, 2-1 and 3-6) used for sedimentary particles and crystalline carbonates throughout this manual. Therefore, it is practice, to conserve consistency, to use the Wentworth scale to describe particle size in igneous rocks. Figure 5-2 shows the traditional grain-size classification and the Wentworth scale as applied to igneous rocks. The latter is recommended for usage.

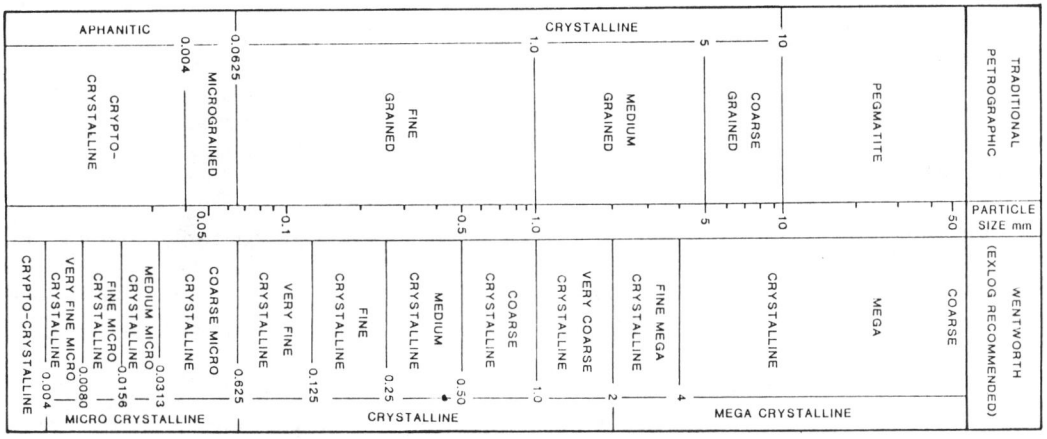

Figure 5-2. Igneous Grain Size Classification

5.12 Texture

In general, igneous rock texture may be separated into three major subgroups:

- holocrystalline: consisting entirely of crystals
- hyaline: consisting of noncrystalline glass
- fragmental: consisting of mixed components from explosive vulcanicity

Under the microscope, igneous rocks have textures defined by the sizes, crystal forms and distribution of their mineral components. There are many variations but, in general, they may be characterized as one of four major types as described in paragraphs 5.13 through 5.16.

5.13 Homeocrystalline: These rocks consist of very uniform-size crystals. They may be fine- to medium-grained and extensive, indicating fairly rapid and uniform cooling. Conversely, the coarsest grained igneous rocks, pegmatites, although occurring in veins or patches, also display a strong uniformity of grain size. The term equigranular is a common synonym for homeocrystalline.

By definition, cryptocrystalline rocks do not have visible crystals under common microscope magnification. However, they may be readily discriminated from hyaline rocks under the microscope by their irregular break, opacity and luster. Hyaline rocks have conchoidal or splintery break, high transparency although generally dark in color, and a vitreous luster.

5.14 Porphyritic: In general, rate of cooling and crystal size are in inverse proportion. Thus the deep-seated large plutonic bodies tend to be coarser crystalline than the smaller shallower volcanic bodies. The occurrence of a porphyritic texture, in which large phenocrysts of relatively uniform size are contained in a fine crystalline ground mass of similar uniform size, may indicate a sudden change in the environment of crystallization such as the movement of partially crystallized magma to a higher level where crystallization is rapidly completed. Thus the phenomenon is characteristic of volcanic rocks and especially minor intrusions. The geologist should not be misled by the presence of large crystals into a false identification.

A special case of porphyritic texture are the hypocrystalline rocks — volcanic glasses containing patches or globules of identifiable crystals. This may be an extreme case of the mechanism given above in which phenocrysts are carried to surface where rapid shock cooling results in the formation of a glass groundmass.

Where the crystalline patches consist not of large single crystals but of microcrystalline aggregates, this explanation seems unlikely. The process of crystal formation probably post-dates solidification of the glass and is due to devitrification. Hyaline rocks are not true solids but "supercooled fluids" and metastable. Over geological time, extremely slow crystallization will occur within them.

5.15 Xenolithic: Intrusive and extrusive rocks may contain fragments of previous sediments or of previously solidified lava (xenoliths) or single crystals (xenocrysts). These may be fresh, show sharp boundaries, readily recognizable by differences in mineralogy and texture from the host rock. Alternatively, they may be partially digested and show only transitional boundaries or appear as patches of color and mineral variation in the rock. In examining such rocks from cuttings, take care to observe evidence of solution and alteration at boundaries of rock fragments and minerals to avoid misidentification of the rock type and emplacement, either as a porphyry or an explosive agglomerate.

5.16 Fragmental: Rocks resulting from explosive vulcanicity can be considered in many ways to be sedimentary rocks in that they consist of clastic fragments transported and deposited at their place of bedding. Although the impetus for transportation is an explosion, both wind and water currents play a part in their eventual sedimentation. Special consideration must be given to the facts that certain of the fragments are fluid or at least plastic at the time of formation, that particle shapes change, and that welding may occur during transport.

In describing the pyroclastic rocks, similar methods and observation should be used as in describing any detrital clastic rock. For convenience, pryoclastic rocks are subdivided into two categories:

- agglomerate: consisting of granule (2 mm) or larger sized particles
- tuff: consisting of sand or smaller-sized particles

Agglomerates are readily recognizable by their total lack of uniformity in any characteristic. Although some size and mass sorting does occur during settling, it is minor. The agglomerate consists of angular fragments (of a wide size range) of sediments and igneous rocks. The sediments may be unaltered or may show metamorphism from contact with the magma. The igneous rocks range from broken air-rounded droplets of glass through fine-grained preexistent lava to coarser-grained intrusive rock. There may also be broken fragments of large single crystals, phenocrysts from the magma. Agglomerates are characterized by the major volcanic rock type present (e.g., "rhyolite agglomerate").

Tuffs may range in grain size from sandstone to claystone and, due to further transport, show better developed rounding, sorting and bedding. They are more difficult to recognize at the wellsite, and are too frequently misidentified. This results in a failure to pick significant markers and may cause confusion when interpreting wireline logs.

Tuffs may be subdivided as follows:

- lithic: those consisting predominately of preexistent rock fragments
- vitric: consisting predominately of whole or broken glass droplets

However, the commonly small particle size, coupled with devitrification, often makes this distinction difficult to make at the wellsite. A more useful subdivision for the petroleum geologist is into "welded" and "nonwelded."

Welded tuffs result from the rapid accumulation and deposition from air of minute liquid droplets and in general coincide with the vitric subdivision above. Nonwelded tuffs are more extensively air- or water-transported and consist of particles which solidified prior to settling. Although generally comparable to lithic tuffs, nonwelded tuffs may contain vitreous fragments and may in fact be the result of reworking of other tuffs or agglomerates.

Welded tuffs are commonly light to dark gray, hard with an angular (but not conchoidal) break. Although no groundmass crystallinity is apparent and the appearance is homogeneous, the rock is translucent to opaque — distinguishing it from volcanic glass. The appearance and character are between that of a smooth waxy shale and a bedded chert, and tuffs are commonly misnamed for one of these extremes. Hardness exceeding that of shale with the absence of shale's fissility or chert's conchoidal break may be a clue. Occasionally, well-defined feldspars may be seen or ferromagnesian porphyroblasts may occur, but these are not universal.

Nonwelded tuffs drill and look like impure argillaceous sandstones. Depending upon the distance of transport, sorting may be relatively good and some rounding of grains may occur. Identification should be made by recognition of the characteristic and relatively fresh mineral assemblage — the sample looks like a crushed igneous rock! Freshness of quartz and ferromagnesian minerals, discrete grains of mica and twinning in relatively fresh feldspars and occasional fragments of whole rock should all indicate the true rock type. Weathering and reworking may destroy much of this character, and the rock may only be identifiable as a detrital sandstone.

Tuffs often occur in sequences, indicating several cycles of vulcanicity. Hard welded tuffs may be overlain by less welded lithic ash with an upper weathered zone, the sequence being repeated several times. When water-deposited, tuffs may be interbedded with or even cemented by chert. The deposition of tuffs increases the silica concentration of the water and promotes increased silica-using faunal populations. This close proximity of bedded cherts and welded tuffs may further complicate identification.

5.17   Grain Form: In addition to the actual size of crystals in an igneous rock, the form of the crystals and boundaries between crystals may be of genetic significance and must be described.

In igneous petrology a large number of terms are used which are unnecessary to, and may involve observations unavailable to, the wellsite geologist. For this reason three basic terms, with modifications where necessary, are used in wellsite descriptions as discussed in paragraphs 5.18 through 5.20.

5.18   Euhedral Granular: A rock in which the majority of minerals are bound by crystal faces is said to be euhedral granular. Other terms used in petrology which are effective synonyms are idiomorphic, automorphic panidiomorphic and lamprophyric (Figure 5-3a).

a) EUHEDRAL GRANULAR b) SUBHEDRAL GRANULAR c) ANHEDRAL GRANULAR

Figure 5-3. Grain Form and Rock Texture

Under conditions of moderate cooling rates (for example, in a small but relatively deep intrusion), the normal sequence of crystallization is overtaken by falling temperatures. Crystals of many species grow simultaneously and result in a euhedral texture and comparatively uniform grain size.

5.19 Subhedral Granular: A rock consisting of minerals with subhedral form only, or more commonly of minerals with a mixture of euhedral, subhedral, and anhedral forms, is referred to as subhedral granular. Synonyms are hypidiomorphic, hypantomorphic, and granitic (Figure 5-3b).

Such texture is typical of slow, uniform cooling in a stable magma and hence of large plutonic bodies. Minerals with higher temperature stability form first and grow largest in euhedral form in the fluid magma. Later-forming minerals are smaller and must form compromise boundaries between preexistent minerals. Reaction with the cooling magma may result in replacement, intergrowth and inclusion of early grown minerals, resulting in partial loss of euhedral form.

A common variation of this texture seen in volcanic rocks consists of a strong interlocking framework of euhedral or subhedral feldspars. The spaces within the framework may be filled with anhedral ferromagnesian minerals (interstitial texture) or glass (intersertal texture).

5.20 Anhedral Granular: If almost all constituent minerals in an igneous rock are anhedral in form, the rock is anhedral granular. Synonyms include xenomorphic, allotriomorphic and aplitic (Figure 5-3c).

Anhedral granular texture is indicative of rapid simultaneous growth and is seen in late-stage plutonic crystallization. Due to their cooling history, the complexity of texture and mineralogy of such rocks may be compared to that of the bittern salts formed in evaporative bodies.

Shallow and extrusive volcanic rocks, due to their variable and episodic cooling history, often exhibit a range of textures often within a single rock. Commonly, the euhedral phenocrysts will be bound in a subhedral or anhedral ground mass. Interstices within the ground mass may contain noncrystalline glass.

5.21 Mineralogy

5.22 Quartz: As stated in paragraph 5.10, the presence or absence of free quartz is a major criterion in classifying an igneous rock. It is essential in an acid rock, it may occur as an accessory in an intermediate or basic rock, and never occurs in an ultrabasic rock.

5.23 Feldspars: These are present in all igneous rocks and, with the exception of the ultrabasic rocks, they are often the major component. Along with quartz and the feldspathoids (see paragraph 5.24), they are the major light-colored minerals found in rocks.

In proceeding further with a rock classification, it is useful to determine the composition of feldspars present. The feldspars rich in sodium and potassium (orthoclase, albite and oligoclase) are essential minerals in acid and in many alkaline intermediate rocks. Sodic plagioclase (andesine) is typical of intermediate rocks, and calcic plagioclase (labradorite, bytownite and anorthite) is typical of basic and ultrabasic rocks (although only in small quantities in anorthite).

Working with cuttings and without access to thin-sectioning equipment, the task of determining feldspar type is difficult. However, if time allows, an attempt should be made. Complete identification is not possible without petrographic equipment, and the geologist at the wellsite must be satisfied with separating the light-colored minerals into quartz, the alkaline (sodium and potassium) orthoclase feldspars and the calc alkaline (sodium and calcium) plagioclase feldspars.

Feldspars are colorless with a vitreous luster when fresh. However, alteration is so common that rarely do they appear this way in ancient rocks. Feldspars are generally seen to be semitranslucent to opaque, pale colored and having a dull porcelaneous luster. Even when weathered, the euhedral or subhedral laths of feldspar are normally visible. Although fresh feldspars have a hardness similar to quartz, alteration usually leaves a softer crumbly surface or edge to the grains.

Plagioclase is commonly grayish-white and very rarely has a reddish coloration. Orthoclase almost always has a reddish color (due to inclusion of ferric oxide) and may range from flesh-tinted to brick red. It is this coloration which in general gives acid rocks a pink color and intermediate and basic rocks a gray color. This rule is not conclusive since plagioclase may rarely be pink and orthoclase rarely gray.

Sample quality and grain size usually prevent the observation of twinning in crystals. However, should suitable cleavage surfaces occur (especially in porphyries), careful inspection should be made. If delicate striations occur, indicating albite twinning, plagioclase is confirmed. The reverse (that no twinning confirms orthoclase) is not necessarily true since twinning is commonly so fine as to be invisible even in plagioclase.

Orthoclase is insoluble in dilute hydrochloric acid. Plagioclase is slowly soluble, becoming increasingly so with increasing calcium content. Solution is not total but leaves a cloudy suspension of colloidal silica.

Obviously, none of these tests are totally conclusive. However, as a general rule, if two feldspar species occur in a rock (as is common), one pink and one gray-white, it can be assumed that the pink species is orthoclase. If any trace of albite twinning is seen in the gray-white species, it may be confirmed as plagioclase.

5.24 Feldspathoids: As the name implies, these are chemically similar to feldspars but poorer in silica. They are comparatively rare but their presence is of major diagnostic importance. Feldspathoids form only from a magma insufficiently saturated in silica to produce feldspars. For this reason they never occur in the presence of free quartz. The most abundant feldspathoids are nepheline and leucite. Sodalite, noselite and hauynite are less common.

Nepheline is very similar to quartz, occurring in irregular grains with a white, occasionally a smokey or reddish coloration and a vitreous or oily luster. Unlike quartz it is readily soluble in dilute acid, leaving a residue of colloidal silica. It is distinguished from quartz by this solubility and by its occurrence in substantial quantities in otherwise dark-colored intermediate rocks.

Leucite forms gray-white, vitreous trapezohedra similar in form to garnet. It is brittle with a conchoidal fracture and occurs only in extrusive rocks. It dissolves in dilute acid, leaving no residue.

Sodalite occurs as bright to dark blue irregular grains in intermediate rocks. It has a strong fluorescence under ultraviolet light and, like nepheline, is readily soluble in dilute acid. Hauynite and noselite are similar to sodalite and cannot be readily distinguished from it at the wellsite.

5.25 Ferromagnesian Minerals: In general, the minerals poor in iron and magnesium are light in color, low in density, and most common in acid rocks. These are quartz, feldspars, feldspathoids, and muscovite (type) micas. Those rich in iron and magnesium, the ferromagnesian or mafic minerals, are dark in color, high in density and most common in basic or ultrabasic rocks.

The presence of dark minerals, without identification, is itself an important key to rock type. The most common ferromagnesian minerals occurring in igneous rocks are biotite (type) micas, pyroxenes, amphiboles, and olivines.

Biotite is the most common dark mineral in acid and intermediate rocks. It occurs only in trace amounts in basic and ultrabasic rocks. Pyroxenes are the major ferromagnesian minerals in ultrabasic and basic rocks, less common in intermediate rocks and rare in acid rocks. Hornblende is typical of intermediate, but accessory amounts do occur in all other rock types.

Using the tools available at the wellsite, further discrimination of the type of ferromagnesian minerals present beyond the groups listed is inconclusive and serves little further diagnostic purpose. Should it be desired and if time allows, refer to a standard mineralogy text.

5.26 ROCK NAME

There is no standard terminology for naming igneous rocks. Names derive from mining and quarrying terms, type, localities, or (occasionally) predominant mineralogy. Hundreds of rock names are in existence, and at least a hundred of these are used regularly. Figure 5-4 is a guide to identifying and naming the more common rock types. While hopefully sufficient for the purposes of the wellsite geologist in petroleum exploration, the list is not complete and includes some generalizations which may be unacceptable to the specialist igneous petrologist.

5.27 METAMORPHIC ROCKS

Like igneous rocks, the occurrence of metamorphic rocks in an exploration borehole is commonly a sign that economic basement has been reached. Unlike igneous rocks, however, metamorphic texture and mineralogy may develop progressively over several hundreds of feet of drilling. Without careful examination by the logging geologist and recognition of the subtle changes in mineralogy and texture, many hundreds of feet of hole may be drilled beyond the point at which abandonment should have occurred.

Where change from sedimentary to metamorphic rock is transitional, which is usually the case, even the most experienced geologist will require time (and drilled footage) to recognize and confirm the event. The geologist should be alert to the possibility of entering a metamorphic zone and should always look for the signs. Experience shows that the majority of footage cut of metamorphic rocks is of a low metamorphic grade which was not recognized as such simply because the geologists involved in the drilling operation did not consider the possibility! In exploration geology, an open mind is the geologist's most important tool.

5.28 CLASSIFICATION

The multiplicity of parents, processes and chemistry of metamorphism leads to a complicated and often conjectural classification system. The purpose of the following paragraphs is to assist the geologist in recognizing metamorphic rocks, and no intensive discussion of classification is given. For this, refer to a standard petrology or mineralogy text.

5.29 Facies

Traditionally, metamorphic rocks are divided into the seven distinct facies groups originally defined by Eskola, with some later modifications. The facies groups are characteristic mineral assemblages dependent upon temperature and pressure. The definition of the facies groups assumes that a mineralogical equilibrium state characteristic of temperature and pressure is reached in the rock so that mineralogy accurately reflects rock history. In practice, this appears to be generally true although complication may result when multiple episodes of metamorphism have been experienced.

A further complication to the diagnostic use of metamorphic facies is that minerals and mineralogical assemblages will have ranges of temperature and pressure in which they are mutually stable. Thus the facies groups do not form a linear series in either temperature or pressure, and may overlap. Finally, the mineralogy of a metamorphic rock, especially at lower grades, will be a function of the mineralogy of the parent rock. At higher grades, the chemistry of the parent rock, its fluids, and any change in material balance, gain or loss during metamorphism will have an effect on the final mineralogy. Figure 5-5 gives the general relationship between the seven major metamorphic facies, and Figure 5-6 shows their characteristic assemblages and occurrence.

5.30 Parent

Characterization of metamorphic rocks by parent can be a simple, useful method for wellsite descriptions. The rock may be described as "meta (igneous or sedimentary rock name)" and followed by a textural and mineralogical description. Figure 5-7 shows the metamorphic mineral products of igneous and sedimentary parents.

A simple classification is by means of the major sedimentary and igneous rock types:

- Pelites: from clays and shales
- Psammites: from pure sandstones
- Semipelites: from impure sandstones, mudstones and silts
- Meta-greywackes: from arkosic sandstones
- Calcareous: from limestones and dolomites
- Acid: from felsic igneous rocks
- Basic: from mafic igneous rocks

Figure 5-7 shows that, at higher grades of metamorphism where chemistry is the controlling influence, these terms are entirely suitable. Certain terms are redundant and others require subdivision. A more concise classification may be made as follows:

- Pelitic: from aluminous pure clay rocks
- Felsic: from quartzose and feldspathic sandstones and acid igneous rocks
- Calcic: from calcitic rocks including impure limestones and dolomites
- Basic: from calcitic rocks containing appreciable quantities of aluminum, magnesium and iron, basic and intermediate rock, and calcitic clays
- Mafic: from ultrabasic rocks and ferromagnesian-rich sediments

Figure 5-8 is an extension of the metamorphic facies concept shown in Figure 5-6. Common mineral assemblages for the five major parents are shown.

5.31 Process

The process whereby metamorphism affected the rock is most significant to the final mineralogy, texture and setting of a metamorphosed section.

EXPLOSIVE (Pyroclastic / Fragmental)	EXTRUSIVE: Hyaline	EXTRUSIVE: Hyaline	EXTRUSIVE: Aphanitic Vesicular	EXTRUSIVE	EXTRUSIVE: Crypto to coarse micro crystalline; Non-to slightly porphyritic	INTRUSIVE: Porphyritic; >50% aphanitic; Crypto to coarse crystalline	INTRUSIVE: Porphyritic; <50% aphanitic; Very fine to mega crystalline	OCCURRENCE / TEXTURE	OCCURRENCE / TEXTURE	OCCURRENCE / TEXTURE	OCCURRENCE / TEXTURE
IV 2mm: (EXTRUSIVE ROCK TYPE) AGGLOMERATE; < 2mm: WELDED / NON-WELDED } (EQUIVALENT EXTRUSIVE ROCK TYPE) TUFF	With spheroidal fractures: PERLITE	OBSIDIAN	PUMICE	FELSITE	RHYOLITE	RHYOLITE PORPHYRY	GRANITE PORPHYRY	>10% ROCK	FREE QUARTZ :	ORTHOCLASE > PLAGIOCLASE	FELDSPAR MORE THAN 10% OF ROCK
					TRACHYTE	TRACHYTE PORPHYRY	SYENITE PORPHYRY	< 10% ROCK			
					PHONOLITE	PHONOLITE PORPHYRY	NEPHELINE SYENITE PORPHYRY	> 10% ROCK	FELDSPATHOIDS:		
	With resinous luster: PITCH STONE				QUARTZ LATITE	QUARTZ LATITE PORPHYRY	QUARTZ MONZONITE PORPHYRY	>10% ROCK	FREE QUARTZ:	ORTHOCLASE::PLAGIOCLASE	
					LATITE	LATITE PORPHYRY	MONZONITE PORPHYRY	< 10% ROCK			
					$Na \geq K$: RHYODACITE	$Na \geq K$: RHYODACITE PORPHYRY	$Na \geq K$: GRANODIORITE PORPHYRY	> 10% ROCK	FREE QUARTZ : SODIC	PLAGIOCLASE :	ORTHOCLASE < PLAGIOCLASE
					$Na \gg K$: DACITE	$Na \gg K$: DACITE PORPHYRY	$Na \gg K$: QUARTZ DIORITE PORPHYRY				
	With spheroidal fractures: PALAGONITE	TACHYLITE	SCORIA	BASALT	$Na \geq K$: TRACHYANDESITE	$Na \geq K$: TRACHYANDESITE PORPHYRY	$Na \geq K$: SYENODIORITE PORPHYRY	<10% ROCK			
					$Na \gg K$: ANDESITE	$Na \gg K$: ANDESITE PORPHYRY	$Na \gg K$: DIORITE PORPHYRY				
					$Ca \geq K$: TRACHYBASALT	$Ca \geq K$: TRACHYBASALT PORPHYRY	DIABASE	ABSENT	OLIVINE: CALCIC		
					$Ca \gg K$: BASALT	$Ca \gg K$: BASALT PORPHYRY					
					BASALT	OLIVINE BASALT PORPHYRY	OLIVINE DIABASE	PRESENT			
					OLIVINE rich: PICRITE						
	Hypocrystalline: VITROPHYRE					With > 50% Mafics: BIOTITE, PYROXENE, HORNBLENDE, OLIVINE } LAMPROPHYRE		ABSENT	OLIVINE :		FELDSPAR LESS THAN 10% OF ROCK
					Nepheline Phenocrysts: NEPHELINITE						
					Leucite Phenocrysts: LEUCITITE			PRESENT			

Figure 5-4. Igneous Rock Classification

MINOR PLUTONIC					MAJOR PLUTONIC				OCCURRENCE			
PORPHYRITIC	EUHEDRAL GRANULAR	VERY FINE TO MEGA CRYSTALLINE	ANHEDRAL GRANULAR	VERY FINE TO COARSE CRYSTALLINE	COARSE MEGA CRYSTALLINE	SUBHEDRAL TO ANHEDRAL GRANULAR	NON-TO SLIGHTLY PORPHYRITIC	VERY COARSE TO COARSE MEGA CRYSTALLINE	TEXTURE			
With >50% Mafics: BIOTITE PYROXENE HORNBLENDE OLIVINE LAMPROPHYRE			GRANITE APLITE	GRANITE PEGMATITE		With Mafics: GRANITE			> 10% ROCK	FREE QUARTZ:	ORTHOCLASE > PLAGIOCLASE	FELDSPAR MORE THAN 10% OF ROCK
						Without Mafics: ALASKITE						
			SYENITE APLITE	SYENITE PEGMATITE		< 50% Mafics: SYENITE			< 10% ROCK			
						> 50% Mafics: SMONKONITE						
			NEPHELINE SYENITE APLITE	NEPHELINE SYENITE PEGMATITE		< 50% Mafics: NEPHELINE SYENITE			> 10% ROCK	FELDSPATHOIDS		
						> 50% Mafics: MALIGNITE						
						QUARTZ MONZONITE			> 10% ROCK	FREE QUARTZ:	ORTHOCLASE = PLAGIOCLASE	
						MONZONITE			< 10% ROCK			
						With Feldspathoids: NEPHELINE MONZONITE						
						Na ≥ K: GRANODIORITE			> 10% ROCK	FREE QUARTZ:	SODIC	ORTHOCLASE < PLAGIOCLASE
						Na >> K: QUARTZ DIORITE						
						Na ≥ K: SYENODIORITE			< 10% ROCK			
						Na >> K: DIORITE						
						With Feldspathoids: NEPHELINE DIORITE						
						With Clinopyroxene: GABBRO	Feldspar only: ANORTHOSITE		ABSENT	OLIVINE:	CALCIC	PLAGIOCLASE
						With Orthopyroxene: NORITE						
						Ca ≥ K: SYENOGABBRO	With Feldspathoids: NEPHELINE GABBRO					
						With Clinopyroxene: OLIVINE GABBRO			PRESENT			
						With Orthopyroxene: OLIVINE NORITE						
						With no other Mafics: TROCTOLITE						
						With Pyroxene: PYROXENITE			ABSENT	OLIVINE:		FELDSPAR LESS THAN 10% OF ROCK
						With Hornblende: HORNBLENDITE						
						Olivine only: DUNITE			PRESENT			
						With no other Mafics: PERIDOTITE						

Figure 5-4. Igneous Rock Classification

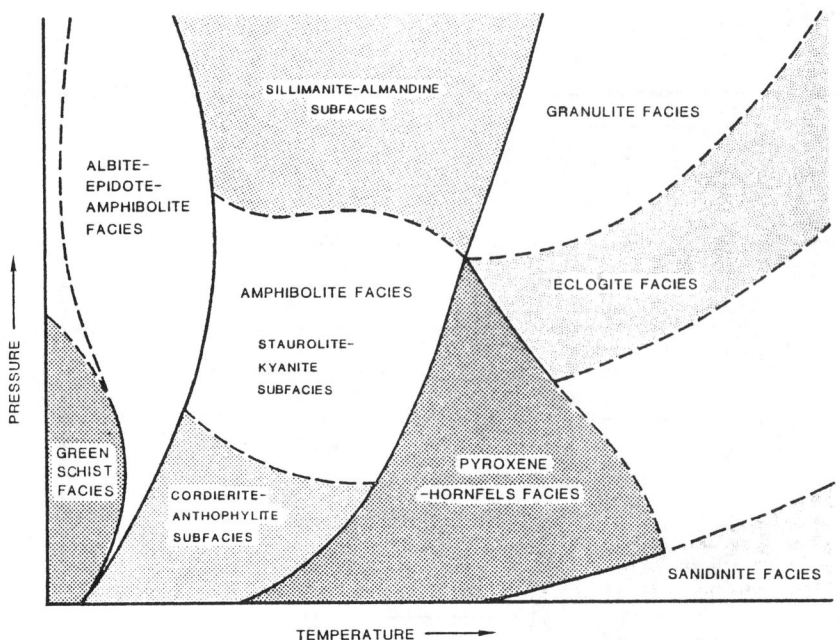

Figure 5-5. Metamorphic Facies

FACIES GROUP	CHARACTERISTIC ASSEMBLAGE	OCCURRENCE		
		TEMPERATURE	PRESSURE	MECHANISM
I GRANULITE	Plagioclase, Diopside, Hypersthene	High	Very High	Regional
II ECLOGITE	Pyrope, Omphacite	High	High?	Partial Melting
III SANIDINITE	Pigeonite, Diopside, Plagioclase	Very High	Low	Igneous Contacts
IV PYROXENE-HORNFELS	Plagioclase, Diopside, Hypersthene	High	Moderate	Contact
V AMPHIBOLITE	(a) Sillimanite, Almandine	Moderate-High	Moderate-High	Regional
	(b) Staurolite, Kyanite	Moderate	Moderate-High	Regional
	(c) Cordierite, Anthopyllite	Moderate	Moderate	Contact
VI ALBITE-EPIDOTE-AMPHIBOLITE	Albite, Epidote, Hornblende	Moderate	Moderate	Regional
VII GREENSCHIST	Albite, Epidote, Chlorite	Low	Moderate	Regional-Hydrothermal

Figure 5-6. Metamorphic Facies: Mineralogy and Occurrence

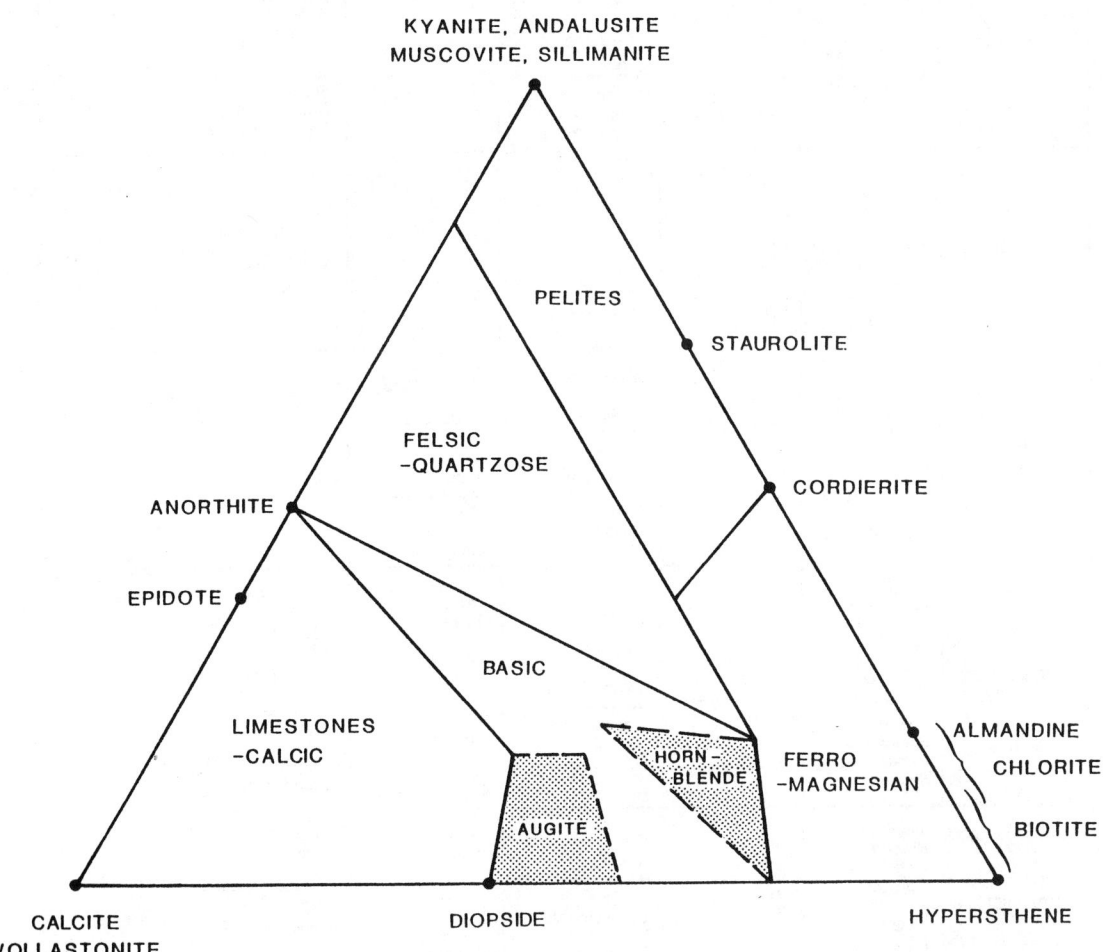

Figure 5-7. Metamorphic Parents and Products

Facies	Pelitic	Felsic	Calcic		Basic	Mafic
			Calcitic	Dolomitic		
I GRANULITE	Quartz - Orthoclase - Plagioclase - Garnet - { Sillimanite, Kyanite }	Quartz - Orthoclase - Plagioclase - Garnet - { Sillimanite, Kyanite, Hypersthene }	Calcite - Plagioclase - Quartz - { Diopside, Scapolite } Plagioclase - Diopside - Hypersthene		Plagioclase - Diopside - Hypersthene - Garnet Olivine - Enstatite - Anthophyllite (-) - Cummingtonite	
II ECLOGITE					Omphacite - Pyrope (-) - Kyanite - Rutile	
III SANIDINITE	Cordierite - Mullite - { Spinel, Glass } Anorthite - Corundum - Spinel	Tridymite - Anorthite - Glass	Sparrite - { Calcite, Larnite } Wollastonite - { Larnite, Quartz }	Melilite - Monticellite - Diopside Melilite - Merwinite - Spurrite Monticellite - Calcite (-) - Periclase		
IV PYROXENE HORNFELS	Biotite - Orthoclase - Quartz - { Andalusite, Cordierite, Sillimanite } Andalusite - Corundum - Spinel	Quartz - Orthoclase - Plagioclase - Biotite	(Calcite -) Wollastonite (-) Grossularite (-) Idocrase - Quartz Anorthite - Scapolite - Grossularite	Calcite - Periclase - (Brucite -) { Forsterite, Humite } Calcite - Diopside - { Forsterite, Wollastonite }	Plagioclase - Diopside - Hypersthene Hypersthene - Cordierite Forsterite - Spinel (-) - Diopside - Hypersthene	
V AMPHIBOLITE	Quartz - Biotite - Muscovite - (Garnet -) { Plagioclase, Staurolite, Kyanite, Sillimanite }	Quartz - Plagioclase - Biotite - Muscovite	Calcite - Quartz Calcite - Diopside (-) { Grossularite, Forsterite }	Calcite - Forsterite - { Brucite, Diopside } Hornblende - Hedenbergite	Plagioclase - Hornblende - Diopside Anthophyllite - (Cummingtonite -) { Biotite, Cordierite }	
VI ALBITE - EPIDOTE - AMPHIBOLITE	Muscovite - Quartz - Biotite - Albite - (Garnet -) (Chloritoid)	Quartz - Albite - Epidote - (Muscovite- Biotite-) Microcline	Calcite - Quartz (-) Tremolite	Dolomite - { Tremolite, Forsterite, Calcite, Talc }	Albite - Epidote - { Hornblende, Actinolite }	Antigonite Chlorite - Actinolite Anthophyllite
VII GREENSCHIST	Quartz - Muscovite - (Albite -) { Biotite, Chlorite, Chloritoid }	Quartz - Albite - Epidote - (Muscovite -) (Microcline)	Calcite - Quartz - { Dolomite, Epidote, Tremolite }		Albite - Chlorite - Epidote (-) { Calcite, Actinolite, Glaurophane }	Carbonate - { Talc (-), Antigorite } Actinolite

Figure 5-8. Common Mineral Assemblages

5.32 Contact: Metamorphism resulting from the intrusion of a hot magmatic body and its associated volatiles is limited in extent. In the case of minor volcanic intrusives, sills and dykes, it may be as little as a few inches. Even the largest plutonic bodies rarely have contact "aureoles" of more than a few hundred feet.

Thickness of metamorphosed sediments may therefore give an estimate of the size of an igneous body on approach (if the body is not vertical or steeply inclined; in which case a false indication may result by drilling parallel to the body).

5.33 Autometamorphism: Digestion of phenocrysts and xenoliths or recrystallization during cooling of an igneous body is a form of metamorphism. Drilling into a small intrusion via a contact aureole and igneous rock contaminated with sedimentary xenoliths may give the impression, when viewed in cuttings, that increasing grades of regional metamorphism are being seen. The speed of transition and the presence of samples of uncontaminated volcanic rock may assist in correcting this false impression.

5.34 Dynamic: In zones of intense deformation and dislocation (for example, thrusts), severe mechanical crushing of rock takes place. Crushed rocks of this type, cataclastics, show major physical but minor mineralogical changes, and the zone is limited to the rocks within the plane of earth movement.

Where regional heating is associated with movement or results from friction, true metamorphism often of a severe stage may occur. The effect, however, is strictly limited, and adjacent sediments are affected only a little.

5.35 Regional: Metamorphism on a major scale combines temperature, pressure, earth movement and plutonism. Such regional metamorphism is always of major extent and, when encountered, indicates that economic basement is reached.

5.36 DESCRIPTION

As stated in paragraph 5.27, failure to recognize metamorphic rocks is commonly due to the failure to expect them. A systematic examination of cuttings samples <u>without the preconception that they are sediments</u> will yield sufficient evidence of their true nature.

5.37 Texture

The texture of a metamorphic rock is a unique product of its mineralogy and the conditions of metamorphism. Even if a complex mineral assemblage is not identifiable at the wellsite, a combination of identifiable minerals and texture should allow the rock to be characterized.

5.38 Cataclastic: Localized zones of severe rock stress may result in fragmentation of the rock and formation of a cataclastic breccia (see paragraph 5.34). The matrix between the cataclastic fragments will consist of a structureless, pulverized rock flour, mylonite.

Cataclastics are the product of localized stress, possibly with minor frictional heating. Little identifiable mineralogical change occurs in the rock. In coarser cataclastic rocks, original minerals retained in place in the rock fragments may show stress damage, shearing or bending.

5.39 Flaser: More intense dynamic metamorphism of the rock will result in physical deformation of the cataclastic fragments and in coarse-grained rocks of original grains and minerals. Streaks and streamlined blebs of rock and mineral fragments will be dispersed in a banded mylonite matrix.

Where the parent was a coarse-grained igneous rock, prophyroblasts of feldspar will remain but will be deformed into streamlined "eye-shaped" masses. This is often referred to as augen texture.

5.40 Mylonitic: The eventual product of dynamic metamorphism is a compact mass of pulverized rock fragments. This rock is dense, featureless, and, when indurated, chert-like. Larger samples may show banding and flow-like structures.

5.41 Hornfelsic: Under conditions of moderate to high temperature and pressure but in absence of an isotropic stress, mineral growth occurs randomly, controlled only by the chemistry of the system. A regular mosaic of minerals results with no preferred crystallographic orientation.

Since crystallization is in a solid medium, crystal boundaries will not result from order of crystallization as in an igneous rock, but from the characteristics of the minerals themselves. Hence quartz, feldspars, cordierite, calcite, etc., will commonly occur in a euhedral crystalloblastic form, whereas rutile, sphene, kyanite, staurolite, and garnet will tend to form anhedral, idioblastic, compromise boundaries. Other minerals will be crystalloblastic or idioblastic, depending upon their association with these two extremes.

Large crystals, porphyroblasts, may occur within the uniform ground mass. These are commonly those minerals with idioblastic tendencies, e.g. garnet, which will readily grow into available space without retention of crystal form. Crystalloblastic minerals may form porphyroblasts where the rock chemistry overwhelmingly favors their formation.

5.42 Granoblastic: Rocks having an originally strong granular texture will tend to recrystallize metamorphically in this form. The result is a compact, anhedral granular rock with irregular, sutured grain boundaries and no interstitial growth. Quartzite and marble are typical of this texture. Where impure, some segregation and lineation of micaceous minerals may result in a weak semischistose texture visible in large samples.

5.43 Slaty: Shales and claystones in which there is strong lineation will tend to recrystallize in a way which reinforces this texture. Shale fissility will be replaced by the stronger slaty cleavage. Commonly there will be no visible mineralogical change in the rock but high temperatures may result in the formation of isolated porphyroblasts.

5.44 Phyllitic: Further metamorphism of slaty rocks results in an increase in grain size and lineation. Commonly the majority of grains are submicroscopic, but a strong micaceous luster develops and a lustrous sheen is most evident at cleavage surfaces. Thermal metamorphism of mylonites may result in induration and incipient lineation. Such rocks, phyllonites, are in small samples indistinguishable from phyllites. They are, however, of far more restricted extent, and their occurrence in narrow shear zones should allow recognition.

5.45 Schistose: Strong lineation of the rock both in mineral growth direction and segregation of minerals is called schistosity. This may result as a development of an already lineated rock or may occur in originally homogeneous rocks due to stress. Schistosity may develop parallel to bedding planes or to surfaces of movement (shear planes). Alternatively, it may form perpendicular to a compressive stress in regional metamorphism. Where both strong bedding and a stress are present during metamorphism, the rock may develop schistosity in two nonparallel planes.

Schistose rocks have strong mineralogical (and hence color and texture) banding. Micaceous and tabluar minerals are common and are usually well enough developed to be visible and identifiable.

5.46 Gneissose: At high grades of regional metamorphism the development of coarse granular quartz and feldspar (in excess of 2 mm) result in the destruction of most early metamorphic texture. However, relics of schistosity, in the form of rough banding and streaking or augen texture, remain to confirm the rock's metamorphic character. Characteristic mineral assemblages and relationships (Figure 5-8 and paragraph 5.41) also allow the rock to be distinguished from an igneous rock.

5.47 Rock Name

Classification of metamorphic rocks requires a combination of mineralogy, macro- and microtexture. The occurrence of episodic or multiple metamorphism may further complicate identification. Figure 5-9 is a general guide to identifying the major metamorphic rock types on the basis of evidence available at the wellsite. It is neither a complete nor an intensive classification, but should prove sufficient to the purposes of the logging geologist in identifying and naming metamorphic rocks.

Texture	Mineral Assemblage		Parent	Metamorphism	Rock Name
	Major	Accessories			
Cataclastic	Undeformed Rock Fragments		Any Rock Type	DYNAMIC	CATACLASTIC BRECCIA
Flaser	Deformed Rock Fragments	Mylonite			FLASER ROCK
Augen	Feldspar "eyes"	Mylonite, Resistate Minerals	Coarse grained igneous rock		AUGEN GNEISS
Mylonitic	Pulverised Rock		Any Rock Type		MYLONITE
Lineated, Mylonitic	Pulverised Rock	Micas		+ THERMAL	PHYLLONITE
Hornfelsic	Various		Clay & Non-Clay Pelites, Impure Sandstones	CONTACT	HORNFELS
Granoblastic	Garnet or Epidote	Calcite, Quartz, etc.	Limestone, Dolomite		SKARN
	Calcite or Dolomite	Tremolite, Diopside, etc.			MARBLE
	Serpentine, Calcite	CaMg Silicates	Dolomite		OPHICALCITE
	Quartz	Various	Pure Sandstone		QUARTZITE
Slaty	Microcrystalline Chlorite, Micas	Quartz, Resistate Minerals	Shale, Mudstones, Impure Fine Sandstones	LOW GRADE REGIONAL	SLATE
Phyllitic	Very Fine Crystalline Chlorite, Micas				PHYLLITE
Schistose	Micas, Chlorite, Quartz	Feldspars, Epidote, Tourmaline, calcite	Shale, Mudstones, Impure Sandstones, Basic Igneous Rocks	LOW TO MODERATE GRADE REGIONAL	MICA SCHIST
	Graphite, Micas, Quartz	Feldspar Chlorite	Carbonaceous Shale, Mudstones		GRAPHITE SCHIST
	Calcite, Micas, Quartz	Graphite, CaMg Silicates	Argillaceous Limestones Dolomites		CALC-SCHIST
	Chlorite, Epidote, Actinolite	Feldspar, Magnetite	Basic Igneous Rocks, Ferruginous Mudstones		GREENSCHIST
	Talc	Magnetite, Mg Silicates, Carbonates	Mafic Igneous Rocks		TALC SCHIST
	Garnet, Micas	Feldspar, Quartz, Hornblende, etc	Any Rock Type		GARNET SCHIST
	Staurolite, Micas	Kyanite, Garnet Quartz	Shales, Mudstones, Impure Fine Sandstones	MODERATE TO HIGH GRADE REGIONAL	STAUROLITE SCHIST
	Sillimanite	Quartz, Micas, Garnet			SILLIMANITE SCHIST
	Hornblende, Feldspar	Garnet, Quartz, Magnetite	Mafic Igneous Rocks and Ferruginous Sediments		HORNBLENDE SCHIST
Gneissose	Feldspar, Quartz	Micas, hornblende Tourmaline, Garnet	Felsic Igneous Rocks, Pure Sandstones, Shale		GNEISS
Gneissose/ Granoblastic	Hornblende, Plagioclase	Garnet, Epidote, Micas	Basic Igneous Rocks, Ferruginous/Calcic Sediments.	HIGH GRADE REGIONAL	AMPHIBOLITE
	Feldspar, Quartz or Pyroxene	Garnet, Kyanite, Tourmaline			GRANULITE
	Garnet, Pyroxene	Kyanite, Rutile	Any Rock Type		ECLOGITE
Veined	Feldspar, Quartz	Biotite, Hornblende		REGIONAL/ PLUTONIC	MIGMATITE

Figure 5-9. Metamorphic Rock Classification

APPENDIX A
FORMATION EVALUATION LOG SYMBOLS

INTRODUCTION

The Formation Evaluation Log usually contains a cuttings lithology column produced with typewriter symbols and an interpreted lithology column which is drafted by hand. The following tables illustrate the recommended symbols to be used for this log.

TYPEWRITER SYMBOLS

These are produced using the logging unit typewriter. One symbol represents ten percent of the sample. The tables specify whether upper or lower case is to be used.

In some cases the symbol must be modified by typing a second typewriter symbol. For example, calcarenite requires a limestone "brick" symbol followed by the sand "dot" symbol. In such cases checkmarks are shown on the table for both characters (see Figure A-1).

In other cases the symbol must be modified by hand after typing. For example, coal requires a limestone "brick" symbol which is then filled in with black ink or pencil. In these cases a checkmark is shown in the "modified" column of the table (see Figure A-2).

DRAFTED SYMBOLS

These are produced by hand, in ink, based upon an overall interpretation of the in-situ formation. Mixed lithologies are represented by mixtures of symbols in the appropriate combination.

DETRITAL ROCKS

The most commonly encountered detrital rocks and accessories rocks are shown in Figure A-1. Although the term "marl" is not recommended for use, symbols are shown for use when necessary. The symbol consists of a horizontal shale "dash" with a vertical short bar at its mid-point.

Remember that when special log symbols are used they must be identified, in their typewritten and hand drafted form, on the log heading.

LITHOLOGY SYMBOLS - DETRITAL

NAME	TYPEWRITER				DRAFTED SYMBOL
	CASE U	CASE L	MODI-FIED	SYMBOL	
CLAY					⟩ ⟩ ⟩
CLAYSTONE		✓		— — — — —	⟩ — ⟩ / — ⟩
SHALE		✓		— — — — —	— — — / — —
SILT & MUDSTONE		✓		═ ═ ═ ═ ═	~ ~ ~ / ~ ~
SANDY SILTSTONE		✓		-·-·-·-·-	~··~ / ···~
SAND & SANDSTONE		✓		· · · · ·	· · · / · · ·
GRAVEL & CONGLOMERATE	✓			o o o o o	o o o / o o
TUFF	✓		✓	⋏⋏⋏⋏⋏	⋏ ⋏ ⋏ / ⋏ ⋏
CHERT	✓			∧∧∧∧∧	△ △ △ / △ △
BIOCLAST	✓			)))))	~ ~ ~ / ~ ~
MARL		✓	✓	⊥⊥⊥⊥⊥	⊥ ⊥ ⊥ / ⊥ ⊥

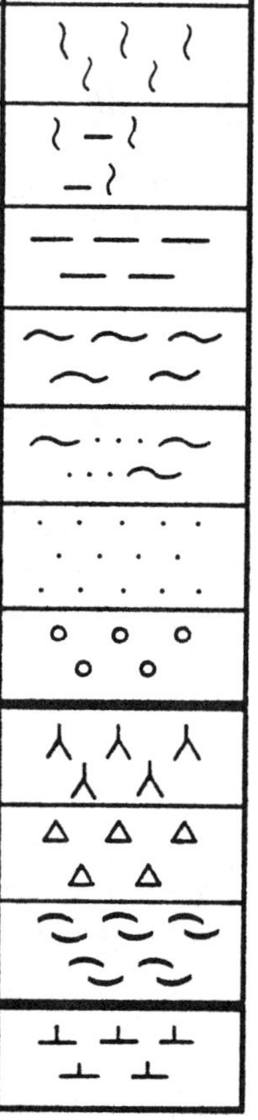

Figure A-1

LITHOLOGY SYMBOLS - ACCESSORIES

NAME	TYPEWRITER				DRAFTED SYMBOL
	CASE		MODIFIED	SYMBOL	
	U	L			
GLAUCONITE		✓		G G G G G	 ～～～ ～～
PYRITE		✓		P P P P P	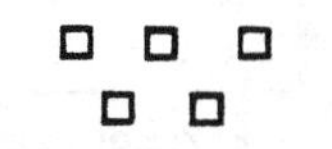 ▫ ▫ ▫ ▫ ▫
HAEMATITE		✓		H H H H H	 ▪ ▪ ▪ ▪ ▪
PHOSPHATES		✓	✓	P P P P P	P P P P P
FELDSPARS		✓		S S S S S	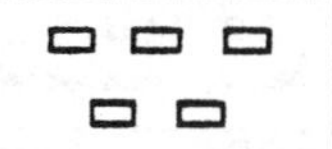 ▫ ▫ ▫ ▫ ▫
MICAS		✓		V V V V V	 ⌵ ⌵ ⌵ ⌵ ⌵
LIGNITE	✓		✓	▬ ▬ ▬ ▬ ▬	

Figure A-2

CARBONATE ROCKS

Figure A-3 shows the standard symbols for carbonate rocks and fossils.

Where more refinement is required, specialized symbols for rock and fossil type are also suggested.

Symbols for marl and chalk are shown, although use of these terms in discouraged. The chalk symbol is similar to limestone but has double vertical bars.

LITHOLOGY SYMBOLS - CARBONATES

NAME	TYPEWRITER				DRAFTED SYMBOL
	CASE U	CASE L	MODIFIED	SYMBOL	
LIMESTONE	✓			▢▢▢▢▢	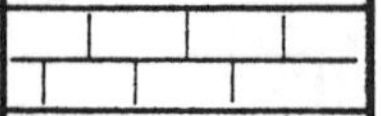
DOLOMITE		✓		Z Z Z Z Z	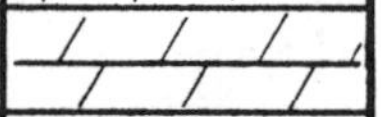
MACROFOSSILS	✓			)))))	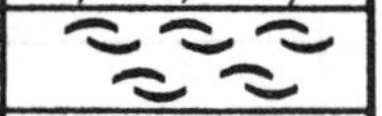
MICROFOSSILS		✓		F F F F F	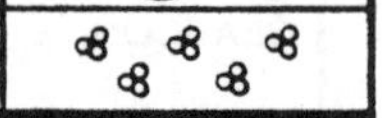
CALCIRUDITE	✓✓			▣▣▣▣▣	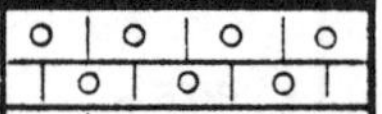
CALCARENITE	✓	✓		⊡⊡⊡⊡⊡	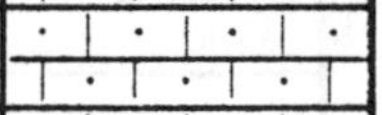
CALCILUTITE	✓			▢▢▢▢▢	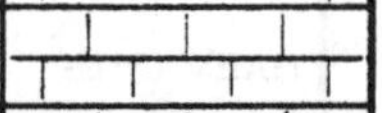
DOLORUDITE		✓		Z Z Z Z Z	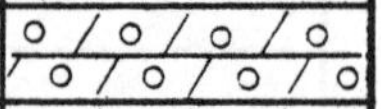
DOLARENITE		✓		Z Z Z Z Z	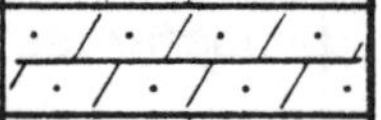
DOLOLUTITE		✓		Z Z Z Z Z	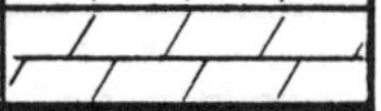
BRYOZOA		✓		F F F F F	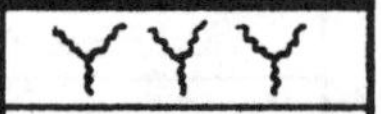
CORALS		✓		F F F F F	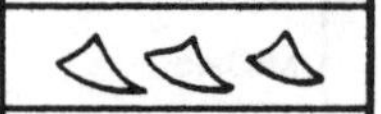
SPONGES		✓		F F F F F	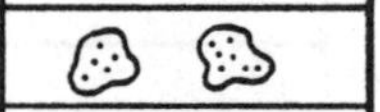
SPINES		✓		F F F F F	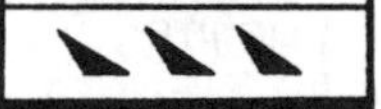
MARL		✓	✓	⊥⊥⊥⊥⊥	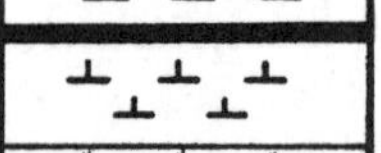
CHALK	✓	✓		▢▢▢▢▢	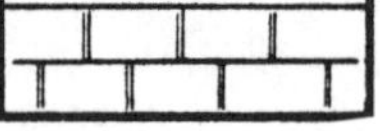

Figure A-3

CHEMICAL ROCKS

The most commonly encountered chemical rocks are shown in Figure A-4. Finer discrimination of the salts and minerals shown is rarely possible with wellsite tools, and it is not recommended that the logging geologist should attempt to do so.

LITHOLOGY SYMBOLS – CHEMICAL ROCKS

NAME	TYPEWRITER				DRAFTED SYMBOL
	CASE U	CASE L	MODI-FIED	SYMBOL	
LIMESTONE	✓			▫▫▫▫▫	(brick pattern)
DOLOMITE		✓		Z Z Z Z Z	(diagonal brick pattern)
ANHYDRITE & GYPSUM		✓		/ / / / /	(diagonal lines)
POTASH SALTS		✓		K K K K K	K K K K K
POLYHALITE	✓			⅄⅄⅄⅄⅄	(cross-hatch pattern)
HALITE	✓			⌐⌐⌐⌐⌐	# # # # #
COAL	✓		✓	■■■■■	(solid black)
CHERT	✓			∧∧∧∧∧	△ △ △ △ △
PHOSPHATES		✓	✓	P P P P P	P P P P P
HAEMATITE & OXIDES		✓		H H H H H	■ ■ ■ ■ ■
PYRITE & SULFIDES		✓		P P P P P	▫ ▫ ▫ ▫ ▫

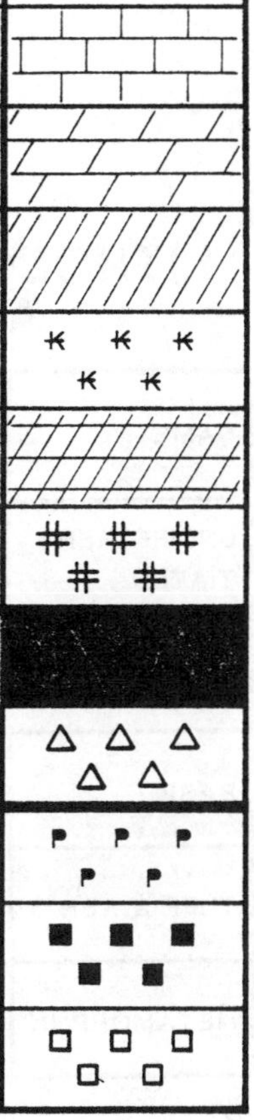

Figure A-4

IGNEOUS AND METAMORPHIC ROCKS

Figure A-5 shows the most commonly used symbols. The rarity of igneous and metamorphic rocks in petroleum exploration and the difficulty of identifying them using available wellsite methods removes the necessity for a wider range of symbols.

In geothermal exploration, igneous and metamorphic rocks are much more significant and demanding of study, requiring specialized techniques and logging procedures.

LITHOLOGY SYMBOLS - IGNEOUS AND METAMORPHIC

NAME	TYPEWRITER				DRAFTED SYMBOL
	CASE U	CASE L	MODI-FIED	SYMBOL	
UNDIFFEREN-TIATED		✓		x x x x x	x x x / x x / x x x
ACIDIC		✓		x x x x x	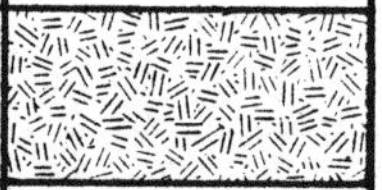 (stippled/dashed pattern)
BASIC		✓	✓	+ + + + +	+ + + / + + / + + +
TUFF & ASH	✓		✓	ʎ ʎ ʎ ʎ ʎ	ʎ ʎ ʎ ʎ / ʎ ʎ ʎ
METAMORPHIC		✓		M M M M	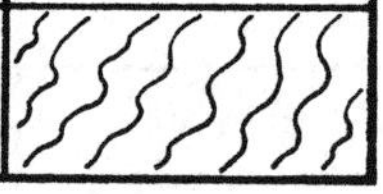 (wavy lines)

Figure A-5

REFERENCES

Anderson, G., 1975, Coring and Core Analysis Handbook, Petroleum Publishing Co.

Asquith, G. B., 1979, Subsurface Carbonate Depositional Models: A Concise Review, The Petroleum Publishing Co.

Blatt, H., G. V. Middleton and R. C. Murray, 1972, Origin of Sedimentary Rocks, Prentice Hall.

Buehler, E. J., 1948, The Use of Peels in Carbonate Petrology, Journal of Sedimentary Petrology, v. 18, p. 71.

Chilingar, G. V., 1956, Use of Ca/Mg Ratio in Porosity Studies, AAPG Bulletin, v. 40, n. 10,

Choquette, P. W., and L. C. Pray, 1970, Geologic Nomenclature and Classification of Porosity in Sedimentary Carbonates, AAPG Bulletin, v. 54, n. 2.

Clementz, D. M., G. J. Demaison and A. R. Daly, 1979, Wellsite Geochemistry by Programmed Pyrolysis, Offshore Technology Conference Proceedings, OTC 3410.

Cowie, I., 1978, Oil and Gas Evaluation Procedures, Exploration Logging, Windsor.

Crossley, A. R., 1979, Some Notes on Lithology Descriptions, Exploration Logging, Singapore.

Deer, W. A., R. A. Howie and J. Zussman, 1967, An Introduction to the Rock Forming Minerals, Longmans.

Dunham, R. J., 1962, Classification of Carbonate Rocks According to Texture, AAPG Memoir, n. 1.

EXLOG, 1985, Field Geologist's Training Guide, IHRDC Press, Boston.

Exploration Logging, 1979, Hydrogen Sulfide Detection: Occurrence and Hazards, MS-3016.

EXLOG, 1985, Mud Logging: Principles and Interpretations, IHRDC Press, Boston.

EXLOG, 1985, Theory and Evaluation of Formation Pressures: A Pressure Detection Reference Handbook, IHRDC Press, Boston.

Folk, R. L., 1959, Practical Petrographic Classification of Limestones, AAPG Bulletin, v. 43, n. 1.

Goldsmith, L. H., 1969, Concentration of Potash Salts in Saline Basins, AAPG Bulletin, v. 53, n. 4.

Greiner, H. R., 1956, Methyl Dolomite of Northeastern Alberta: Middle Devonian Reef Formation, AAPG Bulletin, v. 40, n. 10.

Hobson, G. D., and E. N. Tiratsoo, 1975, Introduction to Petroleum Geology, Scientific Press.

Hopkins, E. A., 1967, Factors Affecting Cuttings Removal During Rotary Drilling, Journal of Petroleum Technology, v. 19, n. 6.

Lane, D. W., 1962, Improved Acetate Peel Technique, Journal of Sedimentary Petrology, v. 32, n. 4.

Leighton, M. W., and C. Pendexter, 1962, Carbonate Rock Types, AAPG Memoir, n. 1.

Low, J. W., 1951, Examination of Well Cuttings, Quarterly of the Colorado School of Mines, v. 46, n. 4.

Maher, J. C., 1959, Logging Drill Cuttings, Guide Book VIII, Oklahoma Geological Survey.

McNeal, 1959, Lithologic Analysis of Sedimentary Rocks, AAPG Bulletin, v. 43, n. 4.

Peterson, J. A. et al, 1965, Sedimentary History and Economic Geology of San Juan Basin, AAPG Bulletin, v. 49, n. 11.

Pettijohn, F. J., 1956, Sedimentary Rocks, Harper.

Pirsson, L. V., and A. Knopf, 1947, Rocks and Rock Minerals, Wiley.

Plumley, W. J. et al, 1962, Energy Index for Limestone Interpretation and Classification, AAPG Memoir, n. 1.

Powers, R. W., 1962, Arabian Upper Jurassic Carbonate Reservoir Rocks, AAPG Memoir, n. 1.

Read, H. H., and J. Watons, 1962, Introduction to Geology, Volume 1: Principles, Macmillan.

Rogers, E. B., 1971, Sand Control in Oil and Gas Wells — 1, The Oil & Gas Journal, v. 69, n. 44.

Scholle, P. A., 1978, Carbonate Rock Constituents, Textures, Cements and Porosities, AAPG Memoir, n. 27.

Sifferman, T.R. et al, 1973, Drill Cutting Transport in Full Scale Vertical Annuli, S.P.E. 4514, 48th Annual Fall Meeting, S.P.E. of A.I.M.E.

Sloss, L. L., 1969, Evaporite Deposition from Layered Solutions, AAPG Bulletin, v. 53, n. 4.

Wardlaw, N. C., 1979, Pore Systems in Carbonate Rocks and Their Influence on Hydrocarbon Recovery Efficiency, AAPG Continuing Education Course Note Series, n. 11.

Williams, C. E. Jr., and G. H. Bruce, 1950, Carrying Capacity of Drilling Mud, Petroleum Transactions Reprint Series: Number 6: Drilling, S.P.E. of A.I.M.E.

Williams, H. et al, 1954, Petrography, Freeman.

INDEX

Aalenian, vi (tab)
Abnormal pore pressure, 54-58
Accessories, 13
   description of, 42–47, 108–109
   fossils, 47
   minerals, 42
Acetate peels, preparation of, 87–88
Acid rocks, 35
   igneous, 145
   metamorphic, 155
Actinolite, 44(tab)
Agglomerates, 149
Albian, v(tab), vi(tab)
Albion, vii(tab)
Alkaline rocks, 35
Allochems, 78
Allotriomorphic rock, 151
Almandine, 43(tab)
Alpine, v-vi(tab)
Aluminous rocks, 123
Andalusite, 44(tab)
Andradite, 43(tab)
Anhedral crystal, 42
Anhedral granular rock, 151
Anhydrite, 45(tab), 126, 129–133
   barium chloride test for, 130–131
   characteristics of, 130(tab)
   conversion to, 131
Anisian, vi(tab)
Annular velocity, cuttings recovery
   and,, 9
Anthracite, 140
Apatite, 46(tab), 123
Aplitic rock, 151
Applachian, vi(tab)
Aptian, vi(tab)
Aquitanian, v(tab)
Aragonite, 46(tab), 93
Arbuckle, vi(tab)
Arcadian, vii(tab)
Archie limestone classification
   system, 76–78
Ardennian, vii(tab)
Arenaceous rocks, 22
Arenigian, vii(tab)
Arenites, 21, 22
Argillaceous rocks, 22–24
   permeability of, 72
Argillites, 21, 22–24
Argoumian, v(tab)
Argovian, vi(tab)
Arkose, 22
Artinskian, vi(tab)

Ashgillian, vii(tab)
Asphaltites, 141
Astain, v(tab)
Astartian, vi(tab)
Asturian, vi(tab)
Atokan, vi(tab)
Attican, v(tab)
Augite, 44(tab)
Austrian, v(tab)
Authunian, vi(tab)
Autometamorphism, 161
Automorphic rock, 150–151
Auversian, v(tab)
Avonian, vii(tab)
Bajocian, vi(tab)
Barite, 46(tab)
Barremian, vi(tab)
Bartonian, v(tab)
Basic igneous rock, 145, 155
Bathovian, vi(tab)
Bauxite, 123
Bedding, 30–33
   convolute, 31
   cross, 31
   current, 31, 32–33
   graded, 31, 32
   laminated, 31, 32
   massive, 33
   medium, 33
   platy, 33
   regular, 31–32
   slump, 31, 33
   thick, 33
   very thin, 33
Bedoulian, vi(tab)
Berriasian, vi(tab)
Bioclasts, 95–96
Bioliths, 97
Biolite, 43(tab), 153
Bioturbation, 33
Bittern salts, 134–136
Bituminous coal, 139–140
Blackriverian, vii(tab)
Blocky cement, 107
Bog iron-ore, 123
Bone beds, 124
Bononian, vi(tab)
Borax, 126
Boudinage, 33
Boundstone, 77(fig)
Bradfordian, vii(tab)
Breccia, 22
Bretonian, vii(tab)

Bruce, G.H., 1,7
Bruxellian, v(tab)
Burdigalian, v(tab)
Burdsandstein, vi(tab)

Calabrian, v(tab)
Calcarenite, 77(fig)
Calcareous, 155
Calcilutite, 77(fig)
Calcimetry, 38
Calcite, 42, 43(tab), 93–94, 126
    sparry, 80
Calcium salts, 129–133
Calclithites, 79
Caledonian, vii(tab)
Callovian, vi(tab)
Cambrian, vii(tab)
Campanian, v(tab)
Canadian, vii(tab)
Cannels, 137
Caradocian, vii(tab)
Carbonaceous rocks, 137–141
Carbonate detritus, 96
Carbonate Energy Index Classification, 84–85(tab), 86
Carbonate minerals
    aragonite, 93
    calcite, 93–94
    dolomite, 94
    particle type and, 93–94
Carbonate rocks, 75–117
    classification of, 75–86
    description of, 86–117
    evaluation log symbols for, 168
Carbonates, 122, 129
    diagenetic, 91
    non-marine, 119–120
    sedimentary, 90–91
Carboniferous Period, vi–vii(tab)
Carnallite, 126, 136
Carnian, vi(tab)
Carrying capacity, 4(fig)
    density and, 1
Casselian, v(tab)
Cassiterite, 44(tab)
Casts, 33
Cataclastic rocks, 161–162
Catagenesis, 60
Cation Exchange Capacity
    (C.E.C.),, 37–38,
    39(fig), 57–58
Cayugan, vii(tab)
Celestine, 126
Cement
    blocky, 107
    description of, 40–42
    drusy, 107
    vs. matrix, 41–42

    porosity and, 69
    rim,, 107–108
Cementation, 13
    description of, 104–108
    diagenetic, 104
    permeability and, 72
    sedimentary, 104
Cenomanian, v–vi(tab)
Cenozoic Era, v(tab)
Chamosite, 122
Champlainian, vii(tab)
Charmouthian, vi(tab)
Chattian, v(tab)
Chautauguan, vii(tab)
Chazyan, vii(tab)
Chemical rocks, 119–141
    evaluation log symbols for, 169
Chemical sediments, 35
Chesterian, vi(tab)
Chlorite, 43(tab), 50
    derivation of, 35
Choquette, P.W., 112
Chromite, 44(tab)
Cincinnatian, vii(tab)
Clarain, 139
Clastic rocks, 21, 34
    induration of, 40
Clay, 24, 41
    dewatering of, 52–54, 56
    diagenesis of, 49–54
    illitization of, 51, 54
Clay minerals, 35
Clay rocks, source potential of, 59–67
Claystone, 24
Clementz, D.M., 63
Coal, 139–140
Coated Surface, 27
Colemanite, 126
Colloclasts, 96
Color, 13
    changes in, 48
    of detrital rocks, 47–48
    of limestones, 109
Color distribution, 48
Comanchean, vi(tab)
Common red hemitite, 46(tab)
Compaction disequilibrium, 55
Conglomerate rocks, 22
Coniacian, v(tab)
Contact metamorphism, 161
Contaminants, in cuttings sample, 10
Coprolites, 124
Corallian, vi(tab)
Core sampling, 18
Cores, sample preparation of, for limestones, 87–88
Corundum, 44(tab)
Couvinian, vii(tab)

Cracking, 60
Cretaceous Period, v-vi(tab)
Cryptocrystalline rock, 148
Cryptofissile response, 38
Crystal structure, 42
Cubic packing, 67, 68, 69(tab)
Cuttings. *See also* Sample
  contaminants in, 10
  permeability of, 72-73
  sample preparation of, for limestone, 88-90
  thickness-to-diameter ratio of, 8
  velocity of, 1
  washed, 15(fig)
Cuttings recovery, 1-9
  ideal, 1, 9
Cuttings sampling, 9-18
Cuttings transport
  gel strength and, 6-7
  laminar flow and, 3, 5
  particle shape and size and, 8
  pipe rotation and, 7-8
  turbulent flow and, 5

Dacian, v(tab)
Danian, v(tab)
Darcy, 71
Decarboxylating, 60
Density
  cuttings recovery and, 1
Depositional porosity, 113
Desander, 9, 17
Desilter, 9, 17
Desmoinsian, vi(tab)
Detrital rocks, 21-73
  classification of, 21-24
  description of, 24-73
  evaluation log symbols for, 165-167
  mineralogy of, 34-38
  petroleum significance of, 48-73
Devonian, vii(tab)
Diagenesis, 39-40
  clay, 49-54
Diagenetic carbonates, 91
Diagenetic cementation, 104
Diagenetic mineralization, 105
Diagenetic recrystallization, 105
Diantian,, vii(tab)
Diapir, 134
Djulfian, vi(tab)
Dogger, vi(tab)
Dolomite, 41, 94, 126
Dolomite growth, selectivity of, 102
Dolomitization, of limestones, 101-102
Domerian, vi(tab)
Dordonian, v(tab)
Doubledeck shale shaker, 12(fig)
Downtonian, vii(tab)

Dresbachian, vii(tab)
Dried sample, sampling procedures for, 15
Drilling fluids
  carrying capacity of, 1, 4(fig)
  density of, 1
  viscosity of, 3-5
Drillipipe, rotation of, 1-8, 9
Drusty cement, 107
Dull surface, 27
Dunham limestone classification system, 76
Dynamic metamorphism, 161
Edenian, vii(tab)
Eifelian, vii(tab)
Emsian, vii(tab)
Enstatite, 45(tab)
Eocene, v(tab)
Eogenetic porosity, 113-114
Epidote, 45(tab)
Epitaxial growth, 108
Equigranular rock, 148
Erian, vii(tab)
Erzgebirgian, v(tab)
Etched surface, 27
Etching, of sample, 87
Euhedral crystal, 42
Euhedral granular rock, 150-151
Euxinic basin, 59
Evaluation log, symbols for, 165-170
Evaporites, 124-136
  calcium salts, 129-133
  carbonates, 139
  sequence of, 124-127
  sulfates, 129-133
Explosive rocks, 144
Extrusive rocks, 144

Fabric
  in detrital rocks, 30
  in limestones, 99
Faceted surface, 27
Facies groups, 154-155, 158(fig)
Famennian, vii(tab)
Feldspars, 152
Feldspathoids, 153
Fenestral porosity, 114
Ferromagnesian minerals, 153
Ferruginous rocks
  carbonates, 122
  hydroxides, 12-123
  oxides, 122-123
  silicates, 122
  sulfides, 123
Firm rock, 40
Fissility, clay diagenesis and, 51
Flame structure, 33
Flaser rocks, 162

Fluorescence, 14
Fluorite, 46(tab)
Folk limestone classification
 system, 78–81
Fossils, 47, 79
Fragmental rocks, 148, 149–150
Franconian, vii(tab)
Frangible rock, 40
Frasnian, vii(tab)
Friable rock, 40
Frosted surface, 27

Galena, 46(tab)
Gamachian, v11(tab)
Gargasian, vi(tab)
Garnet, 43(tab)
Gedinnian, vii(tab)
Gel strength
 cuttings recovery and, 9
 cuttings transport and, 6–7
Geochemical samples, 17
Geothermal gradient, clay diagenesis
 and, 51–52
Givetian, vii(tab)
Glassy surface, 27
Glauberite, 126
Glauconite, 46(tab), 122
Glaucophane, 45(tab)
Gneissose, 163
Goethite, 122
Gothlandian, vii(tab)
Grain form
 anhedral granular, 151
 euhedral granular, 150–151
 in igneous rocks, 150–151
 subhedral granular, 151
Grain shape, 13
Grain size, 13. *See also* Particle size
 in detrital rocks, 25
 in igneous rocks, 147
Grainstone, 77(fig)
Granitic rock, 151
Granoblastic rocks, 162
Greasy surface, 27
Greywacke, 22
Grossular, 43(tab)
Guadalupian, vi(tab)
Guano, 124
Gulfian, v–vi(tab)
Gypsum, 45(tab), 126, 129–133
 barium chloride test for, 130–131
 characteristics of, 130(tab)
 formation of, 131

Halite, 126
 formation of, 133–134
 silver nitrate test for, 133
Hard rock, 40

Hardness, 13
 Moh's scale of, 47
Hauterivian, vi(tab)
Hauynite, 153
Helvetian, v(tab)
Hematite, 46(tab),122
Hercynian, vi–vii(tab)
Hettangian, vi(tab)
Hexagonal packing, 67
Holocene, v(tab)
Holocrystalline rock, 148
Homeocrystalline rock, 148
Hornblende, 44(tab), 153
Hornfelsic rocks, 162
Humic kerogens, 60
Hyaline rock, 148
Hydration, water of, 50
Hydrocarbon generation, conditions
 for, 61
Hydrocarbon shows, 73
Hydrocarbons
 hydrogen/carbon ratio in, 60–61
 solid, 140–141
Hydrogenating, 60
Hydrogrossular, 44(tab)
Hydrolysate rocks, 21, 35–38
 induration of, 40
Hydroxides, 122–123
Hygroclastic response, 38
Hygrofissile response, 38
Hygroturgid response, 38
"Hypabyssal" rocks, 144
Hypantomorphic rock, 151
Hypidiomorphic rock, 151

Idiomorphic rock, 150–151
Igneous rocks, 143–154
 acid, 145
 basic, 145
 classification of, 143–144,
  156–157(tab)
 color classification of, 146–147
 description of, 145–154
 evaluation log symbols for, 170
 grain size of, 147
 intermediate, 145
 mineralogy of, 152–153
 names of, 154
 silica percentage of, 145–147
 texture of, 148–151
 ultrabasic, 145
Illite, 35, 49–50
Illitization, 51–52
Ilmenite, 44(tab)
Inclusions, 13
Induration, 13
 of carbonate rocks, 104
 of detrital rocks, 39–40

Intact material, 97
Interlayer water, 50–51
Intermediate igneous rock, 145
Interstitial water, 50–51
Intraclasts, 79
Intrusive rocks, 144
Iron ore,
  bog, 123
  specular, 46(tab)
Iron oxides, 41
Jointing, clay diagenesis and, 51
Jurassic Period, vi(tab)

Kainite, 126, 136
Kaolinite, 43(tab), 50
  derivation of, 35
Kazanian, vi(tab)
Kernite, 126
Kerogen, 60
  Type I, 61
  Type II, 61
  Type III, 61
Kieserite, 126, 136
Kimmerian, vi(tab)
Kimmeridgian, vi(tab)
Kinderhookian, vii(tab)
Kungurian, vi(tab)
Kyanite, 45(tab)

Lacustrine deposits, of non-marine
  carbonates, 120
Landinian, vi(tab)
Laminar flow, cuttings transport
  and, 3, 5
Lamprophyric rock, 150–151
Landenian, v(tab)
Langbeinite, 126
Laramide, v(tab)
Laterite, 122, 123
Lattorfian, v(tab)
Leighton and Pendexter limestone
  classification system, 83
Leonardian, vi(tab)
Leucite, 153
Leuper, vi(tab)
Levantian, v(tab)
Lias, vi(tab)
Ligerian, v(tab)
Lignite, 139
Limestone categories
  Type I, 80
  Type II, 80
  Type III, 81
  Type IV, 81
Limestone classification
  Archie, 76–78
  Dunham, 76
  Folk, 78–82
  Leighton and Pendexter, 83
  Pettijohn, 76
  Plumley, 86
Limestone cores, sample preparation
  of, 87–88
Limestone cuttings, sample preparation
  of, 88–90
Limestones
  classification of, 76–86
  depositional porosity of, 113
  description of, 86–117
  dolomitization of, 101–102
  eogenetic porosity of, 113–114
  fenestral porosity of, 114
  mesogenetic porosity of, 114
  permeability of, 116–117
  petroleum significance of, 109–117
  porosity of, 109–116
  predepostional porosity of, 113
  primary porosity of, 111
  recrystallization of, 100–101
  sample preparation, 87–90
  secondary porosity of, 111–113
  solution of, 103–104
  telogenetic porosity of, 114–115
Limonite, 44(tab), 123
Lipid, 60
Litharenite, 22
Lithic tuffs, 149
Lithification, 39–40
Lithology, 13
Lladelian, vii(tab)
Llandoverian, vii(tab)
Llanvirnian, vii(tab)
Logging Program, 16
Lower Paleozoic Period, vii(tab)
Ludlovian, vii(tab)
Lusitanian, vi(tab)
Luster, 13
  surface texture and, 27
Lutetian, v(tab)
Lutites, 21
Maestrichtian, v(tab)
Magnesite, 126
Magnesium salts, 135
Magnetite, 44(tab), 122
Malm, vi(tab)
Matrix, 13
  vs. cement, 41–42
  description of, 42
  particle size and, 25
Matrix porosity, 78
Maysvillian, vii(tab)
Meotian, v(tab)
Meramecian, vii(tab)
Mesogenetic porosity, 114
Mesozoic Era, v–vi(tab)
Meta-greywackes, 155

Metamorphic facies, 158(fig)
Metamorphic rocks, 154–164
  basic, 155
  calcic, 155
  classification of, 154–161, 164(tab)
  description of, 161–163
  evaluation log symbols for, 170
  facies groups of, 154–155
  felsic, 155
  mafic, 155
  names of, 163
  parents of, 155. 159(fig)
  pelitic, 155
  processes of, 155–161
  texture of, 161–163
Metamorphism
  contact, 161
  dynamic, 161
  regional, 161
Micaceous hemitite, 46(tab)
Micrite, 80
Microcrystalline calcite ooze, 80
Mid-Devonian, vii(tab)
Millidarcy, 71
Mineral wax, 140–141
Mineralization
  diagenetic, 105
  permeability and, 72
  secondary, 104
Mineralogy
  of carbonate rocks, 100–104
  of igneous rock, 152–153
Minerals, 42
Miocene, v(tab)
Mirabilite, 126
Mississippian, vii(tab)
Missourian, vi(tab)
Mohawkian, vii(tab)
Moh's scale of hardness, 47
Moldic porosity, 101
Monazite, 45(tab)
Monomodal sorting, 27
Montian, v(tab)
Montmorillonite, 35, 49–50
Morrowan, vi(tab)
Moscovian, vi(tab)
Moulds, 33
Mud
  sampling of, 17
  defined, 24
Mudstone, 24, 77(fig)
Mudweight, cuttings recovery and, 1, 9
Muschelkalk, vi(tab)
Muscovite, 44(tab)
Mylonitic rocks, 162
Namurian, vi(tab)
Natron, 126
Neocene, v(tab)

Nepheline, 153
Nevadian, vi(tab)
Niagaran, vii(tab)
Non-marine carbonates
  lacustrine deposits of, 120
  precipitation of, 119
Nonswelling response, 38
Nonwelded tuffs, 149–150
Norian, vi(tab)
Noselite, 153

Ochoan, vi(tab)
Oil staining, 14
Oligocene, v(tab)
Oligomict rocks, 34
Oolites, 79, 97
Opal, 43(tab)
Ordovician, v(tab)
Oregonian, v(tab)
Organic material
  conversion of, to hydrocarbons, 60
  Type IV, 61
Organic-rich sediments, 59–60
Orthochemical sediment, 80
Orthoclase, 43(tab), 152
Orthoquartzite, 22
Osagean, vi(tab)
Overpressure cap rock, 56
Overpressure seal, 56
Oxfordian, vi(tab)
Oxides, 122–123

Packing geometry, 67–69
  permeability and, 71
Packstone, 77(fig)
Palaeocene, v(tab)
Palaeogene, v(tab)
Palatine, vi(tab)
Paleozoic Era, vi–vii(tab)
Palynology samples, 17
Panidiomorphic rock, 150–151
Pannonian, v(tab)
Parents, of metamorphic rocks, 155, 159(fig)
Particle alignment
  porosity and, 69
Particle shape
  cuttings transport and, 8
  of detrital rocks, 26
  of limestones, 98
  permeability and, 72
Particle size. See also Grain size
  of carbonate rocks, 91–93
  cuttings transport and, 8
  of detrital rocks, 25
  permeability and, 71
  shale shaker and, 10, 11(fig)

Particle type
  in carbonate minerals, 93–97
  clastic debris, 95–96
  intact material, 97
  origin of, 94–95
Pasadenian, v(tab)
Pearly surface, 27
Peat, 138
Pelites, 21, 155
Pelitic sediments, 48
Pellets, 79, 97
Penjabian, vi(tab)
Pennsylvanian, vi–vii(tab)
Permeability
  description of, 70–73
  of limestones, 116–117
Permian, vi(tab)
Petroleum significance
  in detrital rocks, 48–73
  of limestones, 109–117
Pettijohn limestone classification system, 76
Pfalzian, vi(tab)
Phenocrysts, 148
Phosphatic rocks, 123–124
Phosphorite, 123
Phyllitic rocks, 163
Piacezian, v(tab)
Pipe dope, 14
Pipe rotation
  cuttings recovery and, 9
  cuttings transport and, 7–8
Pisolites, 79
Pitted surface, 27
Plagioclase, 43(tab), 152
Plastic rock, 40
Pleistocene, v(tab)
Pliensbachian, vi(tab)
Pliocene, v(tab)
Plumley limestone classification system, 86
Plutonic series, 143
Polished surface, 27
Polyhalite, 126, 136
Polymict rocks, 34
Polymodal sorting, 27
Pontian, v(tab)
Pore pressure, abnormal, 54–58
Porosity
  cement and, 69
  depositional, 113
  description of, 67–70, 116
  eogenetic, 113–114
  fenestral, 114
  of limestones, 109–116
  mesogenetic, 114
  particle alignment and, 69(fig)
  predepositional, 113

primary, 111
secondary, 109–113
sorting and, 69(fig)
telogenetic, 114–115
Porphyritic rock, 148
Portlandian, vi(tab)
Potash salts, 135
Potassium ions, in illilization, 51
Potassium salts, 135
Powers, R.W., 52
Pray, L.C., 112
Precambrian, vii(tab)
Precipitation, of non-marine carbonates, 119
Predepositional porosity, 113
Priabonian, v(tab)
Primary porosity, 111
Psammites, 21, 155
Psephites, 21
Pterocerian, vi(tab)
Purbeckian, vi(tab)
Pyrenean, v(tab)
Pyrite, 41, 44(tab), 123, 126
Pyroanalyzer, 63
Pyrobitumens, 141
Pyroclastic rocks, 149
Pyrolosis, 63–67
Pyrolosis log
  example of, 66(fig)
  interpretation of, 65–67
Pyrope, 43(tab)
Pyroxenes, 153

Quartz, 43(tab), 152
Quartzarenite, 22
Quaternary Period, v(tab)

Rauracian, vi(tab)
Recrystallization
  diagenetic, 105
  of limestones, 100–101
  recognition of, 105–106
  sedimentary, 104
Red hemitite, 46(tab)
Regional metamorphism, 161
Relief, surface texture and, 27
Resistate rocks, 21, 35
  induration of, 40
Rhaetian, vi(tab)
Rheology, 1
Rhodanian, v(tab)
Rhombohedral packing, 67
Rhumba shaker, 10, 12(fig)
Richmondian, vii(tab)
Rim cement, 107–108
Rock class, 90–91
Rock classification, 18, 19(tab)

Rock description
    detrital rocks, 24–73
    interpretive remarks in, 25
Rock-Eval pyroanalyzer, 63
Rock type, 13
Rocks
    aluminous, 123
    arenaceous, 22
    argillaceous, 22–24
    carbonaceous, 137–141
    carbonate, 75–117, 168
    chemical, 119–141, 169
    detrital, 21–73, 165–167
    ferruginous, 121–123
    igneous, 143–154, 170
    metamorphic, 154–164, 170
    phosphatic, 123–124
    rudaceous, 21–22
    saline, 124–136
    siliceous, 120–121
Rotliegendes, vi(tab)
Roundness, particle shape and, 26
Rudaceous rocks, 21–22
Rudites, 21
Rupelian, v(tab)
Rutile, 45(tab)

Saalian, vi(tab)
Sabkha, 129
Sakmarian, vi(tab)
Saline rocks, 124–128
Salopian, vii(tab)
Salts
    bittern, 134–136
    calcium, 129–133
    potash, 135
Samartian, v(tab)
Sample. *See also* Cutting
    combined, 17–18
    description of, 73
    dried, 15
    geochemical, 17
    labeling of, 16–17
    limestone, 87–90
    lithology of, 13
    mud, 17
    palynology, 17
    shipping of, 16–17
    unwashed, 10–13
    washed, 13–15
Sampling
    of core, 18
    procedure, 10–17
    requirements, 16
Sand, 22
Sandstone, 22
Sannoisian, v(tab)
Santonian, v(tab)

Sapropelic kerogens, 60
Sardinian, vii(tab)
Savian, v(tab)
Saxonian, vi(tab)
Scale of hardness, 47
Schistose, 163
Secondary mineralization, 104
    permeability and, 72
Secondary porosity, 111–113
Sedimentary carbonates, 90–91
Sedimentary cementation, 104
Sedimentary recrystallization, 104
Sedimentary rocks, accessories
    in, 42–46(tab)
Semipelites, 155
Senecan, vii(tab)
Senonian, v–vi(tab)
Sequanian, vi(tab)
Shale, 23–24
Shale Factor Test, 37–38
Shale shaker, 9–17
    doubledeck, 12(fig)
    rhumba, 10, 11(fig)
    sampling from, 10–17
Siderite, 41, 122
Siegenian, vii(tab)
Silica, 41
Silicates, 122
Siliceous rocks
    inorganic, 121
    organic, 120–121
Silky surface, 27
Siltstone, 24
Silurian, vii(tab)
Sinemurian, vi(tab)
Skythian, vi(tab)
Slaty rocks, 162
Slip velocity, 3–5
Slippage, 1
Slow drill log, sampling of, 16
Sodalite, 153
Soft rock, 40
Solid hydrocarbons, 140
Soluable rock, 40
Solution, of limestones, 103–104
Sooty surface, 27
Sorting, 1, 13
    comparative, 29(fig)
    description of, 27–30, 98–99
    mechanical analysis of, 29(fig)
    monomodal, 27
    polymodal, 27
    porosity and, 69(fig)
Sparnacian, v(tab)
Sparry calcite, 80
Specular iron ore, 46(tab)
Spessartine, 43(tab)
Sphalerite, 46(tab)

Sphericity, particle shape and, 26
Spot plate, 14
Staurolite, 45(tab)
Stephanian, vi(tab)
Stratigraphic column, v–vii(tab)
Striated surface, 27
Structure, 13
　description of, 30–33, 99–100
Styrian, v(tab)
Sub-Hercynian, v(tab)
Subhedral crystal, 42
Subhedral granular rock, 151
Sucrosic structure, of limestones, 102
Sudetian, vi(tab)
Sulfates, 41, 129–133
Sulfides, 123
Supratidal ground waters, 129
Surface texture, description of, 27, 98
Sylvite, 126, 136

T(max), 65
Taconian, vii(tab)
Teleogenetic porosity, 114–115
Terrigenous detritus, 96
Terrigenous rocks, 21. See also Detrital
　rocks
Tertiary Period, v(tab)
Texture
　in igneous rock, 148
　of metamorphic rocks, 161–163
Thanetian, v(tab)
Thenardite, 126
Thermonatrite, 126
Thickness-to-diameter ratio, of
　cuttings, 8
Tillite, 22
Time stratigraphic column, v–vii(tab)
Titanite, 45(tab)
Tithonian, vi(tab)
Toarcian, vi(tab)
Tongrian, v(tab)
Topaz, 45(tab)
Tortonian, v(tab)
Tourmaline, 45(tab)
Tremadocian, vii(tab)
Trempealeauian, vii(tab)
Tretonian, vii(tab)
Triassic Period, vi(tab)
Trona, 126
Tuffs, 149–150
Turbulent flow, cuttings transport and, 5
Turonian, v(tab), vi(tab)
Type I kerogen, 61
Type II kerogen, 61
Type III kerogen, 61
Type IV organic material, 61

Ulexite, 126

Ulsterian, vii(tab)
Ultrabasic igneous rock, 145
Ultraviolet-light box, 14
Unconsolidated rock, 40
Unwashed samples, 10–13
Upper Paleozoic Period, vi–vii(tab)
Uvarovite, 43(tab)

Valangianian, vi(tab)
Van Krevelen diagram, 60, 61(fig)
Varves, 131
Vermiculite, 50
　derivation of, 35
Vesuvianite idocrase, 46(tab)
Villafranchian, v(tab)
Vindobonian, v(tab)
Virgillan, vi(tab)
Virglorian, vi(tab)
Viscosity, cuttings recovery and, 3–5, 9
Visean, vii(tab)
Visible porosity, 78
Virtrain, 139
Vitreous texture, 27
Vitric tuffs, 149
Vitrinite reflectance, 139
Volcanic association, 144
Vraconian, vi(tab)
Wackestone, 77(fig)
Wallachian, v(tab)
Washed cuttings
　examination of, 15(fig)
　sampling procedures for, 13–15
Water
　of hydration, 50
　interlayer, 50–51
　interstitial, 50–51
Waxy surface, 27
Wealden, vi(tab)
Welded tuffs, 149–150
Wentworth classification, 21
Wentworth particle, 10, 11(fig)
Werfenian, vi(tab)
Westphalian, vi(tab)
Wettability test, 38
Wichita, vi(tab)
Williams, C.E., Jr., 1, 7
Wolfcampian, vi(tab)

Xenocrysts, 149
Xenolithic rock, 149
Xenomorphic rock, 141

Ypresian, v(tab)

Zircon, 44(tab)
Zoisite, 46(tab)